SILENCING SCIENCE:
National Security Controls and Scientific Communication

Information Management, Policy, and Services
Charles R. McClure and Peter Hernon, Editors

Organizational Decision Making and Information Use
Mairead Browne
Library Performance Accountability and Responsiveness: Essays in Honor of Ernest R. DeProspo
Charles C. Curran and F. William Summers
Gatekeepers in Ethnolinguistic Society
Cheryl Metoyer-Duran
Curriculum Initiative: An Agenda and Strategy for Library Media Programs
Michael B. Eisenberg and Robert E. Berkowitz
Resource Companion to Curriculum Initiative: An Agenda and Strategy for Library Media Programs
Michael B. Eisenberg and Robert E. Berkowitz
Information Problem-Solving: The Big Six Skills Approach to Library & Information Skills Instruction
Michael B. Eisenberg and Robert E. Berkowitz
The Role and Importance of Managing Information for Competitive Positions in Economic Development
Keith Harman
A Practical Guide to Managing Information for Competitive Positioning in Economic Development
Keith Harmon
Librarianship: The Erosion of a Woman's Profession
Roma Harris
Microcomputer Software for Performing Statistical Analysis: A Handbook for Supporting Library Decision Making
Peter Hernon and John V. Richardson (Editors)
Public Access to Government Information, Second Edition
Peter Hernon and Charles R. McClure
Record Management and the Library
Peter Hernon and Candy Schwartz
Statistics: A Component of the Research Process, Revised Edition
Peter Hernon
Statistics for Library Decision Making: A Handbook
Peter Hernon et al
Libraries: Partners in Adult Literacy
Deborah Johnson, Jane Robbins, and Douglas L. Zweizig
Library and Information Science Research: Perspective and Strategies for Improvement
Charles R. McClure and Peter Hernon (Editors)
The National Research and Education Network (NREN)
Charles R. McClure, Ann P. Bishop, Philip Doty, and Howard Rosenbaum
Technology and Library Information Services
Carol Anderson and Robert Hauptman
U.S. Government Information Policies: Views and Perspectives
Charles R. McClure, Peter Hernon, and Harold C. Relyea
U.S. Scientific and Technical Information Policies: Views and Perspectives
Charles R. McClure and Peter Hernon
Depository Library Use of Technology: A Practitioner's Perspective
Jan Swanbeck and Peter Hernon
For Information Specialists
Howard White, Marcia Bates, and Patrick Wilson

SILENCING SCIENCE:
National Security Controls and Scientific Communication

Harold C. Relyea
Congressional Research Service
The Library of Congress

ABLEX PUBLISHING CORPORATION
Norwood, New Jersey

Printed in the United States of America

Library of Congress Cataloging-in-Publication Data

Relyea, Harold.
Silencing science : national security controls and scientific communication / Harold C. Relyea.
p. cm. – (Information management, policy, and services)
Includes bibliographical references and index.
ISBN 1-56750-096-X. – ISBN 1-56750-097-8 (pbk.)
1. Communication in science–United States. 2. Communication of technical information–United States. 3. National security–United States. 4. Science and state–United States. 5. Technology and state–United States. 6. Research–Government policy–United States. I. Title. II. Series.
Q223.R45 1994
338.97306–dc20 93-50697
CIP

Ablex Publishing Corporation
355 Chestnut Street
Norwood, NJ 07648

For My Mother and My Father

CONTENTS

ABOUT THE AUTHOR

Harold C. Relyea is a Specialist in American National Government with the Congressional Research Service of the Library of Congress. An undergraduate of Drew University, he received his doctoral degree in government from The American University. Since joining CRS in 1971, he has produced a number of major studies for Congress, including analyses of the office and powers of the President, executive branch organization and management, congressional oversight, and various aspects of government information policy and practice.

In addition to his CRS duties, Dr. Relyea has authored over 100 articles for scholarly and professional publications in the United States and abroad. His books include *The Presidency and Information Policy* (1981), *Freedom of Information Trends in the Information Age* (1983), *Striking A Balance: National Security and Scientific Freedom* (1985), and *United States Government Information Policies: Views and Perspectives* (1989). He is currently completing a book on the Executive Office of the President.

Dr. Relyea also serves on the editorial boards of *Access Reports*, *Government Information Quarterly*, the *Journal of Media Law and Practice*, and *Presidential Studies Quarterly*. He resides in Falls Church, Virginia, with his daughter and son.

ABOUT THE AUTHOR

PREFACE

From the earliest days of the Republic until the dawn of modern global war, Federal policy has been largely supportive of the tradition in scientific research, particularly campus-based studies, of open communication, free exchange of information, and wide publication. The experience of World War II convinced the Nation's leaders of the value of scientific enterprise for defending the country. The subsequent Cold War, however, seemed to distort this recognition. There appeared to be no limit to the number and kinds of contributions of science that were perceived by Federal officials to be vital to the national security of the United States and, therefore, in need of control to limit their availability. Consequently, despite protests from the American scientific community, various restrictions were placed on traditional scientific communication. Relief from these security controls occurred as Cold War anticommunism gave way to international trade and foreign policy of greater cooperation with the Soviet Union and its Warsaw Pact allies during the 1960s. Nevertheless, the legacy of control authority remained available. As relations with the East bloc deteriorated in the late 1970s, culminating with the end of detente on the occasion of the Soviet invasion of Afghanistan, the old restrictions on scientific communication were slowly reactivated and new controls were also sought. Subsequently, during the first half of the 1980s, American scientists saw a vast increase in national security controls on their professional communication, teaching, and speech.

Many scientists, both for reasons of principle and function, opposed the increasing of controls. To them, the free dissemination of information and open discussion were essential parts of the scientific process. Ideally, in their view, findings should be publicly and generally available and subject to criticism, improvement, and, if warranted, rejection. Inevitably, research and, consequently, scientific progress suffered from such broad restrictions. One only had to look at the condition of science in the Soviet Union to appreciate this point. Some limits were conceded to be necessary, but it was thought that they should take the form of "tall fences around narrow areas" of science.

Moreover, many in the scientific community argued for security by achievement in science, not by concealment. They pointed out that "national security" and "science" were not necessarily antagonistic to one another. In the past, scientists and government leaders had demonstrated a broad appreciation of the national security concept, including not only military applications and preparations, but also economic, cultural, and psychological considerations. For many within the scientific community, national security, in its most meaningful context, was not defined by the military capability of hostile nations alone.

During the first half of the 1980s, scientists appeared to be both uncertain as to a strategy for combatting the onset of increased national security controls and ineffective in obtaining relief from the new restrictions. Federal officials demonstrated that a variety of communications limitations could be invoked and applied, with consequential effects on teaching, including both subject matter and student participation, presentations at professional conferences and symposia, personal and professional writing, research options, and personal and professional associations.

It soon became apparent that the application of such controls was linked to the changing fortunes of national security, as perceived by Federal officials. The considerations involved might be diplomatic, involving changing United States relationships with other countries; military, regarding the defense significance of a scientific discovery or research finding; juridical, concerning changing law, civil liberties conditions, or government autonomy standards; tactical, involving responses to new initiatives by foreign powers to secure sensitive scientific knowledge; political, resulting from the selection of new leaders; or personal, deriving from the ideological perspective of a powerful official. Differences of opinion can and do exist concerning the significance of changing conditions in the world, and they can prompt scientists and government leaders to disagree over the need for national security controls on scientific communication.

Indeed, toward the end of the decade, scientists experienced a mitigation of national security controls and a shift in national security interests as a result of changing circumstances over which they had little influence. Defense officials, who had personally championed greater national security regulation of scientific communication, departed from the Pentagon. Simultaneously, new concerns about the nation's trade position prompted support for policies of greater U.S. competitiveness in overseas markets and a relaxation of national security restraints on many technological goods and services. Then came the *perestroika* reforms of Mikhail Gorbachev, effectively ending the Cold War, largely prompting the democratization of the East bloc, and significantly improving relations between the United States and the Soviet Union.

By the dawn of the next decade, however, another set of changing circumstances was emerging. Once again, the Pentagon began pursuing stringent national security restriction of the foreign availability of certain broad categories of scientific knowledge and technologies concerning, for example, semiconductor materials and microelectronic circuits, software producibility, data fusion, and composite materials. Some other technologies, such as superconductivity and biotechnology materials and processes, were being considered for national security protection in both a military and economic competitiveness context. Furthermore, the deterioration and demise of the Soviet Union and growing antipathy among policymakers for Japanese trade practices were prominent factors among various conditions influencing the reformulation of the national security posture of the United States.

The situation brightened with the arrival of the Clinton Administration. For example, although many aspects of the comprehensive national security position of the United States remained unsettled, President William Clinton, in late September of 1993, announced certain initiatives to boost American computer and telecommunications trade, including the easing of Cold War export restrictions pertaining to such technologies. Nonetheless, the legacy of national security authority for controlling scientific communication remains available and largely undiminished.

The silencing of science in America, for reasons of national security, is the subject of this book. While exploring this difficult and controversial contemporary public policy issue, this study also struggles with the elusive nature of the national security concept, the way national security policy is formulated and expressed, and countervailing protections of speech and publication, including traditional scientific communication in the United States.

The opening chapter examines some concepts that are central to this book—what is being regulated—before considering the significance of professional communication in scientific research, federal government support and regulation of science and technology, and the interrelationship of science, technology, and the national interest. The central regulatory value—national security—is addressed in the next chapter, including its historical origins and Cold War development, means of formulation and expression, legal and policy evolution, and interrelationship with science and technology. A historical review of the emergence of national security restrictions on traditional scientific communication, with emphasis on the period from World War I to contemporary times, is undertaken in the third chapter. Included as well is a review of the current operative control laws, the general nature and purpose of such controls, and countervailing protections of traditional scientific communication in the United States. The fourth and fifth chapters are devoted to recent federal efforts at applying national security controls on scientific communication to university-based scientific research endeavors and to technology and technical data emanating from American industry. Responses from the professional community, industry, the new technologies community, and Congress are considered. The final chapter reviews new and rejuvenated threats to America, in the aftermath of the demise of the Soviet Union, that could prompt a renewed resort to national security controls and the possibilities for realigning national security goals and traditional scientific communication.

Familiarity with and understanding of the issues and events discussed here were facilitated by the author's particular participant observer roles with the Congressional Research Service of the Library of Congress, where he has been a senior staff member for more than two decades, and the Committee on Scientific Freedom and Responsibility of the American Association for the Advancement of Science, where he served as a member for six years.

A note of acknowledgment and appreciation is extended to the Council of Scientific Society Presidents, the Council of Engineering and Scientific Society Executives, the American Association for the Advancement of Science, the American Society of Access Professionals, the National Federation of Abstracting and Information Services, *Government Information Quarterly*, *Transnational Data Report*, *Library Trends*, and *National Forum*, among others, for opportunities to develop assessments and exchange views regarding the issues addressed in this book. Moreover, having observed much of the Corson panel proceedings at the National Academy of Sciences in 1982, it was my privilege to serve as the AAAS liaison to the successor Allen panel in 1985–1986.

Permission to utilize ample quotations from the 1985 report, *Securing Technological Advantage: Balancing Export Controls and Innovation*, was granted by the Center for Strategic and International Studies; from Abraham Lincoln's April 4, 1864, letter to Albert G. Hodges permission was granted by The Abraham Lincoln Association; from the 1981 *Report of the Public Cryptography Study Group* permission was granted by the

American Council on Education; from Harold D. Lasswell's 1950 research study, *National Security and Individual Freedom*, permission was granted by the Committee for Economic Development; from Robert L. Park's article, "The Muzzling of American Science," appearing in the September-October 1986 issue of *Academe*, permission was granted by the American Association of University Professors; and from Richard H. Ullmann's article, "Redefining Security," appearing in the Summer 1983 issue of *International Security*, permission was granted by MIT Press Journals. Quotations are also reprinted with permission from *Scientific Communication and National Security*, copyright 1982; *Global Trends in Computer Technology and Their Impact on Export Control*, copyright 1988; and *Finding Common Ground*, copyright 1991, courtesy of the National Academy Press, Washington, DC. Material appearing in *Science* magazine and other publications copyrighted by the American Association for the Advancement of Science is quoted with the permission of A.A.A.S.

Finally, several individuals contributed in special ways to the development of this book. Among them are Tim Ingram, former staff director of the House Subcommittee on Government Information and Individual Rights, and the late Gerry Sturges, a close personal and professional friend who managed the Subcommittee's prescient 1980 hearings on the government's classification of private ideas. At various times, ideas and insights were provided by Allan Adler, Rosemary Chalk, Jack Crowley, Dick Ehlke, Jim Ferguson, Louis Fisher, Stephen Gould, Ruth Greenstein, Marcel LaFollette, Stuart Macdonald, Dick Meserve, Dorothy Nelkin, Bob Park, Marc Rotenberg, Mitchel Wallerstein, and Vivian Weil. My gratitude to these individuals and to many others who contributed to my understanding of the issues and provided information. To Peter Hernon, a special note of thanks in all of these regards and more.

Falls Church, Virginia
November 1, 1993

1

FRAMING THE CONFLICT

Efforts to exert, for reasons of national security, greater government control over scientific communication by American scientists during the 1980s were largely premised on two general considerations. The first of these concerned international or, more explicitly, East–West relations. As a consequence of continued tensions between the two superpowers, the aggressive aims of Soviet foreign policy, and the threat of Soviet attainment of military superiority over the defense capabilities of the United States, it was in the interest of the United States, some argued, to limit the availability of certain scientific information. This could be accomplished through more stringent application of export controls or—as was done in the aftermath of the Soviet invasion of Afghanistan—the official curtailment of the level of cooperative activity under scientific exchange agreements. The kinds of scientific information to be so restricted was a determination to be made by the government, depending on its interpretation of *national security*.

The second consideration involved what might be called *responsible guardsmanship*. Scientists throughout the private sector of American society—in both the academic and the business-industrial communities—had contributed to a "hemorrhage" of technology to the Communist bloc, it was contended, by openly communicating sensitive scientific information in publications, in classrooms, and at professional conferences. The extent to which such knowledge was "sensitive" in a national security context was a judgment that the government sought to reserve largely to itself. Scientists were urged to exercise restraint and to seek government counsel when in doubt about an open communication of possibly sensitive information. In some fields of research, such as cryptography, formal review procedures were prescribed.

The result was a highly complex controversy having foremost significance for American science as well as vital importance for the nation. Although the dispute subsided, no policy resolution was reached. Consequently, prior to examining the events and issues of the recent altercation, this chapter considers the elements involved. In brief, what was being regulated? What was the conceptual basis for the increased controls? What authorities for control were available, created anew, expanded, or intensified?

SCIENCE AND TECHNOLOGY

Almost four centuries ago, the Dutch scholar Desiderius Erasmus warned that every definition is dangerous. Undoubtedly, there are many reasons for this precaution. Nonetheless, this examination of increased national security controls on traditional scientific communication necessarily begins with a look at the essential concepts involved. Such an undertaking is not without importance for this analysis or merely an academic exercise. The central regulatory value—national security—is a nebulous term of shifting meaning and changing understanding in federal law and public policy. Indeed, its indeterminate nature, contrary to Erasmus' admonition, appears to have contributed significantly to the problems surrounding increased national security controls on scientific communication.

"When I use a word," Humpty Dumpty scornfully told Alice, "it means just what I choose it to mean—neither more nor less." Questioned by her as to the possibilities of giving a single word many different meanings, the haughty eggman replied that the issue was "which is to be master—that is all." National security, as shown in the next chapter, is a master concept of sometimes dangerously unclear meaning.

The object of national security regulation—*science*—although better understood, nonetheless also warrants clarification. In today's world, science comprises a method by which knowledge is gained, including identifying a problem, forming a hypothesis, gathering data, testing the hypothesis, and forming general conclusions.[1]

The concept also embraces the body of knowledge resulting from that process, "the constellation of facts, theories, and methods collected in current texts."[2]

Furthermore, a distinction has long been made between at least two kinds of research—so-called basic and applied studies—in science. In his celebrated 1945 report to the President on a program for postwar scientific research in America, Vannevar Bush described *basic research* as being "performed without thought of practical ends." The purpose was discovery for its own sake. Indeed, it results, wrote Bush, "in general knowledge and an understanding of nature and its laws." Such general knowledge, he pointed out, "provides the means of answering a large number of important practical problems, though it may not give a complete specific answer to any one of them."

By contrast, Bush noted, it is the function of *applied research* "to provide such complete answers" and result in "practical applications" of research knowledge. Linking the concepts, basic research "leads to new knowledge and "provides scientific capital," he explained. "It creates the fund from which the practical applications of knowledge must be drawn."[3] Moreover, the conduct of applied research may take various forms and representations. Some years ago, Lord Rothschild identified 33 different terms for applied research.[4] Although his evaluation reduced these to four subsets, the list, nonetheless, reflects the diverse nature of applied research.

Admittedly, in some areas of scientific study, such as biotechnology, distinctions between basic and applied research are difficult to draw. There also appears to be a degree of confusion as to the differences between basic and applied research and *fundamental research*. Among scientists, fundamental research involves studies of such natural phenomena as the elasticity of polymers, catalysts, and nuclear particle attraction. Although the objective is not knowledge for its own sake, the eventual specific application of fundamental research findings may not be known at the time the research is being performed. Nonetheless, there is some expectation that a practical benefit will be discovered in the near future.[5]

Some recent policy definitions, however, have muddled the meaning of the term. In a May 1984 congressional hearing, the Reagan Administration unveiled a draft policy statement endorsing security classification as the principal mechanism "for control of fundamental research in science and engineering at universities and Federal laboratories."[6] Although the concept of fundamental research was not defined in the draft statement, an Administration representative told House overseers that the term referred to "a scientific endeavor that expands our basic knowledge and understanding of the world and the universe."[7] Later, in September 1985, when the draft statement was formally issued by the President as National Security Decision Directive 189, fundamental research was defined as:

> basic and applied research in science and engineering, the results of which ordinarily are published and shared broadly within the scientific community, as distinguished from proprietary research and from industrial development, design, production, and product utilization, the results of which ordinarily are restricted for proprietary or national security reasons.[8]

Finally, some understanding of *technology* is warranted. Over a decade ago, a historian of science noted the fashionable and influential nature of the term, and regarded it as useful to scientists to obtain funds for their work or status.[9] The observation has been offered that, in its earliest usage, technology "signified mechanical tools and implied machinery of various kinds," and that "the purpose of all technology can be generalized as an attempt to modify in some intended and desired way the relationship or compatibility of man and his environment."[10] Moreover, whereas science seeks to understand reality, technology applies knowledge to "alter and control the material conditions of life" or to "create a reality according to our design."[11] Yet, regarding the relationship of science and technology, the following comment has considerable merit:

> It is hard to distinguish the boundary lines between basic and applied science and technology. The point is that both basic and applied science are a part of technology. Thus, basic science is an information function; and applied science is an information function with a useful purpose in mind; while technology is the development and social use of information. A great deal of technology innovation, over the years, came into being without the aid of science; and conversely, a great deal of the information uncovered by science has not found useful application but is still judged as potentially useful, or as the basis for useful understanding. Also, to the extent that basic science involves measurement or observation employing tools and devices it is dependent on previous technological innovation. Many great scientific advances resulted from this sequence, as for example from the microscope, telescope, spectroscope, etc.[12]

These terms, however, at times have not been used precisely or consistently in law and public policy, or with the same definition or general understanding discussed here. The explanation of fundamental research in NSDD 189 is an apt example. As this situation creates various kinds of confusion, it certainly blurs what is being regulated—processes of information or knowledge transmittal.

COMMUNICATION AND TRANSFER

There are, in fact, at least two kinds of transmittal that became the object of increased national security controls. One of these, *scientific communi-*

cation, may be described as the open conveyance of scientific information or knowledge by one or more individuals to one or more individuals largely for professional purposes. The following characteristics may be ascribed to it. There are two elements involved–scientific information and communication. At a minimum, scientific information is derived by the scientific method and may also include knowledge of practical utilization to produce useful processes, products, or programs for public or private purposes. Thus, basic, applied, and fundamental research are embraced. The communication or transmittal is deliberately open, meaning it is generally accessible and does not have the private character of a personal letter or individual conversation, and may be made through a variety of media–publication, teaching, professional conference participation, and speeches or lectures. Although the individuals who originally produced the scientific information usually initiate its premiere communication, a variety of other persons also are involved in varying ways in the scientific communication process, perhaps as secondary reporters or interpreters of findings or as facilitators of the transmittal effort. Finally, these communications are made largely for professional purposes–independent verification, criticism, improvement, and the advance of knowledge.

The second kind of transmittal, *technology transfer,* may be described as the public conveyance of a technological product (good, service, or information) or process (trade secret) by a person or group to others largely for professional or commercial purposes. The two essential elements involved are the technology as such and its transfer. The transfer or transmittal is public, that is it may be accessible to some extent, or only the event of the transaction is known. However, it does not have the private character of a personal letter or individual conversation, and may be made through the same variety of media identified previously for scientific communication, but also including commercial sales. Further, although many of these transmittals are for the same purposes as scientific communication, many others occur for commerce and trade.

Few American scientists appear to have taken a position of opposing all federal national security constraints on scientific communication and technology transfer. Moreover, a reading of the public press and professional literature suggests there is little current dispute over the federal government's traditional application of national security controls to a few special areas of science and technology involving munitions, implements of war, or innovations of clear military application. The end of the Cold War has not eroded this consensus, and greater public awareness of international terrorism has probably contributed to increased acceptance of such regulation.

To some extent, federal funding of research and development endeavors has induced tolerance of national security controls on their performance and results. There also seems to have been a consensus for some time for minimal national security restrictions so long as voluntary and proprietary constraints served to limit the public availability of scientific and technological knowledge of national security concern to the government. In brief, to the extent that national security restrictions were selective and limited, they were not regarded as a danger to achievement and progress in science and technology (which, arguably, contribute to the security and well-being of the nation).

Controversy arose over whether or not actual conditions warranted or justified increased national security controls, as well as over the breadth of the new restrictions, and their seemingly adverse impact on achievement and progress in science and technology in the United States. This clash of views and related events in the development of the control issue are discussed in subsequent chapters. The focus here is on performance and values affected by increased national security limitations.

OPENNESS IN SCIENCE

The advancement of knowledge, including its clarification, interpretation, diffusion, and evaluation, is the basic function of the scientific community.[13] This is accomplished, in large measure, through a high degree of open communication and free sharing of information. Indeed, the open character of science has proven to be essential for the advancement of knowledge and, concurrently, vital for the detection and elimination of error.[14]

Openness, however, has not always been an attribute of science. In ancient Greek society, for example, secretiveness resulted from esotericism—the belief that only the elect can penetrate the mysteries of science. For the medical community, the *Law* of the Hippocratic Corpus borrowed the language of the mystery religions, saying: "Holy things are revealed only to holy men. Such things must not be made known to the profane until they are initiated into the mysteries of knowledge."[15] Esotericism was also operative among the Pythagoreans who apparently safeguarded their beliefs and thinking zealously to prevent the contamination that would result from these beliefs becoming known to the outside world.[16]

In medieval Europe, science was "conditioned by a rigid social hierarchy and by a political and religious outlook that encouraged withholding the secrets of nature from the 'vulgar'."[17] The practice of alchemy exemplified the issues of secrecy in this period. As an attempt to discover the relationship of humankind to the cosmos, it was scientific; as an attempt to exploit that relationship for human benefit, it was technological. In seeking to transform base metals into gold, to discover a universal cure for disease, and to find a means of indefinitely prolonging life, alchemists were intensely secretive, both to protect their work for personal gain and also to guard against public disapproval and suspicion.[18]

The Middle Ages, however, also witnessed the beginning of some institutions and technology that encouraged openness in science. Poor communications among scholars and scientists were significantly improved with the advent of printing, which generally allowed scientists to reach a broader and more diversified audience. In addition, the isolation of scientists was alleviated somewhat with the rise of new universities, though some innovative thinkers found these institutions unsympathetic to new knowledge that might change the established order.[19]

Renaissance scholars held secular knowledge in higher esteem. Originality in thought, art, and science attracted patrons and prompted diversification. Revelations about the "secrets" of statecraft in Niccolo Machiavelli's *Il Principe* and Martin Luther's criticisms and disclosures of the political endeavors of the papacy suggested that some civil and ecclesiastical limits on human knowledge had broken down.[20] "By the seventeenth century," wrote one authority, "there was a widespread realization that traditional injunctions against forbidden knowledge had resulted in the stagnation of knowledge rather than the protection of valuable secrets."[21]

Openness in science was undoubtedly fostered further by two often intertwined human motivations—prestige and profit. Those unlocking the secrets of nature wanted proper credit for their discoveries. Moreover, they wanted to retain a degree of control over complicated long-term research, as well as some rein on the immediate use of their findings. (Fear of civil and ecclesiastical reprisal was also a consideration.) Consequently, an early stratagem was to publish scientific results in the form of an anagram. This practice established priority of discovery while leaving the author some control over the masked knowledge.[22]

That empirical knowledge could be profitable was clear to artisans and craftsmen of the Middle Ages. Theirs were technological skills—for example, weaving, masonry, metallurgy, or dye making—that, for both economic and social reasons, were carefully protected. This secrecy was reinforced by the guilds that dominated the commercial life of later medieval Europe and firmly entrenched the idea that craft technique constituted private property.[23] As the guild structure broke down in the later Middle Ages, a more open scientific information environment supposedly gained acceptance. Individual craftsmen, it has been contended, recognized the need to keep improving their knowledge in order to continue to compete in the marketplace, and also came to understand that their own skills were the result of cumulative information (heretofore maintained by the guilds).[24]

Another view, however, argues that Renaissance scholars studying the artisan tradition and technical subjects, not the artisans themselves, stressed the public and collaborative character of knowledge.[25] There was almost unanimous agreement among 16th-century writers on technology that knowledge of technical processes was not dependent on chance or the cleverness of inventors or magic, but, rather, derived from understanding the causes of natural phenomena.[26] The open availability of scientific knowledge facilitated both scientific and technological achievement and progress.

An important accommodation between open scientific and technological information, which when put to practical use could result in profit, and the protection of such intellectual property, to prevent plagiarism and consequential economic loss, occurred with the rise of copyright and patent law. The first patent statute, for example, appeared in 1474. Enacted in Venice, the law granted an inventor a 10-year right of exclusive use in exchange for the disclosure of the invention so it would be publicly available when the patent period expired. "More important," one scholar commented, "it recognized the social utility of patents and saw protection of intellectual property as necessary for advancing technological knowledge."[27] Further discussion of intellectual property protection is reserved for later in this chapter.

Another impetus for openness occurred as the organization of science developed, particularly with the rise of scientific societies. The first such bodies, like the Academia die Segreti of Naples founded in 1542, were clothed in secrecy to protect their members and activities from disapproving civil authorities.[28] Serious and more widespread efforts at organizing formal scientific societies in the 17th century, however, were afforded a different and apparently attractive model to follow in this regard. Shortly after Francis Bacon's death, his literary executor published his uncompleted *New Atlantis*, wherein an ideal scientific research institute—the House of Salomon—was described. All natural phenomena were to be examined through observation and experiment; research was to be conducted collegially, with investigators organized hierarchically in terms of subject matter and level of analysis. Secrecy was practiced, but only to prevent intrusions and castigation by the state, not to restrict those laboring within Salomon's House.[29]

When Bacon's model was put into practice, however, some significant modifications developed. The Academie des Sciences of Paris, which began informally shortly after the middle of the 17th century, was concerned not only with the original experiments of its members, but also with the validation of discoveries by others outside its ranks. Priority of discovery and professional recognition soon became problematical as disputes arose between Academie members and outsiders, as well as internally among the members themselves. Consequently, in 1688, the Academie ceased the practice of publishing its members' in-house findings under its collective aegis. By the end of the century, its constitution was revised to facilitate individual research endeavors by Academie members and any pretense of collegial study in the Salomon's House fashion disappeared.[30]

When the Royal Society of London was chartered in 1662, its members pursued their own original investigations from the outset and regarded the Society as a forum for reporting and discussing findings. As a consequence of this activity, and in deference to another Bacon dictum prescribing written recording of knowledge concerning experiments, detailed notes were made of Society deliberations. Moreover, with the appointment of Henry Oldenburg as secretary, an active correspondence with European researchers was initiated that "within the space of only a few years made the Society an international centre for the exchange of scientific and technical information."[31]

Due to a combination of factors—further realization of Bacon's dictum, Oldenburg's ambition, and a desire on the part of many Society members to establish the priority of their discoveries and attain personal recognition for their achievements—the Society inaugurated a publication in 1665: the *Philosophical Transactions*. It served as not only a major source of open scientific information, but also a repository of validated or authenticated knowledge. To realize this latter purpose, the Society called on its members to referee submissions to its publication.[32]

With the development of Baconian scientific societies and scientific publications like the *Philosophical Transactions*, the practice of openness became institutionalized. Furthermore, increased professionalization was realized as well. By the 19th century, critical sensitivity to not only the substance of new discoveries (originality, authenticity, and duplicability), but also their manner of disclosure (publication, report circulation, and demonstration) was an established characteristic of science and technology professionalism. In brief, research findings had to be presented in accordance with certain expectations (publicly, thoroughly, and intelligibly) before their contribution as new knowledge would be considered.[33] This attitude has largely prevailed in the 20th century.

SCIENTIFIC FREEDOM

As this historical discussion of openness in science illustrates, investigators of the mysteries of nature, from at least the dawn of Christianity to the 18th century, were well aware that their freedom to conduct research and report resulting discoveries was subject to limitation (and sometimes severe restriction) by the ecclesiastical and civil authorities of the day, as well as the baser instincts of society. Fear of reprisal by the church, the state, or the mob was a major reason for cloaking scientific activities in secrecy, even in the idealized world of Salomon's House. However, much had changed, the late Carl Becker has reminded us, by the dawn of the 18th century—"the moment in history when men experienced the first flush and freshness of the idea that man is master of his own fate."[34] Science and technology, of course, had contributed significantly to this conviction. The workings of nature were but expressions of the will of God; to fathom the phenomena of nature was to reveal the beneficent intentions of the Almighty. The discovery of these truths through human reason and accumulated and verifiable knowledge meant there was no need to rely upon official and dogmatic pronouncements of the church and the state.[35]

As science and technology subsequently advanced, however, pronouncements about holy revelations deriving from the analysis of nature were forsaken. Furthermore, as research results provided humankind with power over things and even over other human beings, it became increasingly clear that discoveries could be used indifferently for any purpose, whether good or bad. Scientists were gaining greater freedom to select their research subjects, conduct their studies, and utilize their findings. Ironically, although science and technology flourished in an atmosphere of freedom, their fruits could very well be used to restrict or destroy that freedom.

Scientific freedom prescribes that each scientist has individual autonomy over the selection of research subject matter, professional communication, and critical validation or invalidation of scientific findings. Greater progress, achievement, and knowledge integrity in science and, to some degree, in technology result to the extent that these choices are unrestrained. However, although scientific and technological results owe no allegiance to any particular form of government, science and technology are not pursued in a vacuum. Consequently, constraints have been set on scientific freedom through law as well as official policies and practices. The scientific and technological professions have also imposed restrictions by setting ethical standards and reminding practitioners that there are responsibilities that accompany freedom.

One very important responsibility of scientists is, in fact, to advance knowledge. Biochemist John T. Edsall has argued that scientific freedom is an "acquired right": Scientists are granted an extraordinary measure of freedom in the expectation that knowledge beneficial to society will be yielded. The professional expertise of scientists conveys special respon sibilities. "Scientists," wrote Edsall, "can claim no special rights, other than those possessed by every citizen, except those necessary to fulfill the responsibilities that arise from the possession of special knowledge and of the insight arising from that knowledge."[36]

Giving support to Edsall's "acquired right" description, Harold P. Green, a longtime legal analyst of scientific freedom issues, found no "precedent or legal authority that clearly supports the proposition that there is a constitutionally protected right to pursue knowledge or to engage in scientific inquiry." He contended, however, "that such rights are implicit in the First Amendment freedom of speech and press."[37] The proponents of a restriction on scientific freedom, noted Green, carry a heavy burden of proof.[38] A particularly strong governmental interest in the limitation, such as protecting against a clear and present danger, must be demonstrated,[39] and the constraint must be structured so that it intrudes on constitutionally protected utterances to the minimum extent possible.[40] A different situation, however, arises with experimentation, which is "within the range of action that may enjoy only some, or perhaps very little or no, constitutional protection," in Green's view. The crucial consideration here is the impact on people and the environment: "Where scientific research involves experimentation with human or animal subjects or where it impinges upon the community, it would clearly become subject to regulation."[41]

Modern day governmental restraints on scientific freedom have varied over time in different political settings. Some have been of minor impact, short duration, or otherwise of little concern to the scientific community. By contrast, others recounted here have prompted organized responses. For example, from time to time during the past 40 years, the American Association for the Advancement of Science (AAAS), the world's largest general science society and also the world's largest federation of scientific organizations, and the National Academy of Sciences have created special study panels to examine and report on various scientific freedom concerns. In late December 1947, AAAS established a Special Committee on Civil Liberties for Scientists.[42] It reported the following December. At issue were federal loyalty-security standards and procedures and their impact on scientists. Although it was concluded that both science and democracy were "damaged" by secrecy, the panel also conceded that "there will undoubtedly remain some matters of scientific cognizance which should be kept confidential." Consequently, loyalty-security investigations of federal personnel, including scientists, given knowledge of these matters was warranted in order to assure their reliability.[43] The committee, however, was critical of prevailing standards for determining loyalty that focused on employees' associations rather than their behavior in overt acts. Its report also warned that the failure to confine loyalty determinations to matters of objective proof made federal workers feel insecure "and may be expected to lessen the vigorous intellectual independence which is a prime condition of sound scientific work as it is of an imaginative civil service."[44] Particular concern was expressed about security clearance requirements in the atomic energy program that had been extended to any scientists having personal contact with colleagues serving the Atomic Energy Commission. The panel indicated that this development, if allowed to continue, would prompt many scientists to pursue studies unrelated to the atomic energy field.[45]

The impact of federal loyalty policy and practice on American science was again sufficiently controversial in early 1955 that presidential assistant Sherman Adams was prompted to request guidance from the National Academy of Sciences. A Committee on Loyalty in Relation to Government Support of Unclassified Research was created by Academy president Detlev W. Bronk.[46] Its report was completed in mid-March and released by the White House in early April. By its own account, the committee focused exclusively on "the question of alleged or confirmed disloyalty on the part of scientists, not employed by the Government, engaged in unclassified research under Government sponsorship in private and State universities, hospitals, and laboratories of the country."[47] At issue was the course of action that should be taken when a federal agency found a serious question of loyalty in the case of a scientist being sponsored by the agency to conduct unclassified research. Due to a lack of statutory guidance on the matter, the panel noted that "departments and agencies supporting unclassified research have pursued somewhat varying courses." Its recommendation, in effect, was to do nothing: "The proper objectives of Government in sponsoring basic research will be best served by concentration on scientific competence alone." Because unclassified research had no immediate direct bearing on the security of the United States, the loyalty of those performing it was irrelevant. Federal agencies sponsored unclassified research to serve the public interest by advancing scientific knowledge as quickly as possible. If funds for these endeavors were misused, pertinent criminal laws should be applied. Only by destroying the scientific integrity of the unclassified sponsored research could a disloyal scientist sabotage the public interest. However, the committee counseled, the critical scrutiny of the scientific community serves to protect the sponsoring agency, the nation, and science against this risk.[48]

AAAS also established a study panel in 1955—the Interim Committee on the Social Aspects of Science—to examine the state of science in the United States and its relation to social forces and issues.[49] Among the findings in the panel's 1956 preliminary report were some that had particularly troubling implications for scientific freedom. For example, the committee noted the occurrence of "an unprecedented growth in the scale and intensity of scientific work," but was concerned that this was unbalanced. Applied research and development rather than basic science received the major part of research support, which, moreover, was slanted heavily toward the physical sciences. Also, "a very large part of our total research activities are for military purposes"—approximately 84% of federal research funds being so committed in 1957. There was, as well, a dependence by colleges and universities on federal funds for the greater portion of their research support.[50] In view of these developments, was the selection of research subject matter being adversely affected?

The committee also noted problems in scientific communication. Professional journals were insufficient in number, often slow to publish, and inadequate in their abstracting, indexing, and reviewing efforts. In addition, federal support of scientific research, particularly for military and national security purposes, introduced restrictions on communications (private industry was criticized for this as well).[51]

"At a time when decisive economic, political, and social processes have become profoundly dependent upon science," said the committee, "the discipline has failed to attain its appropriate place in the management of public affairs."[52] Scientists were exhorted to play a more active role in social affairs and to consider how their profession could contribute to society.

Succeeding the Interim Committee and building on its work was the AAAS Committee on Science in the Promotion of Human Welfare. In a 1960 report, it concluded that the scientific community "bears a serious responsibility" to assist in solving problems resulting from "the interactions between the social order and the progress of science," and suggested a program in this regard.[53] The report also noted that "conscious exploitation of science for military advantage continues at an accelerating rate"[54] (perpetuating, presumably, limitations on scientific communication[55]), and seemed to warn against scientists using their professional competence or status to give authority to their personal opinions about public policy.[56] However, it said the scientific community did bear an obligation to provide the citizenry with objective knowledge to assist them in making informed decisions on science-related public issues, and did have special competence to detect and report on incipient problems.[57] Finally, among the questions of immediate importance that the committee identified for the scientific community to address, at least three affected scientific freedom directly: increased military support of research and attendant secrecy restrictions, unbalanced growth in science due to "the absence of any over-all [federal] rationale in the support of science and the overemphasis on projects that give promise of immediate practical results," and possible diminution of the integrity of science if a more socially active scientific community was pressured to relax the traditional rules for the conduct of research.[58]

Each of these issues was addressed again, in a more elaborate way, in a 1965 report by the committee.[59] This report found that secrecy, with all of the problems and difficulties it posed for scientific endeavors, had "become ubiquitous in modern, large-scale science and technology";[60] unbalanced growth in science continued because "agencies which are responsible for the pattern of research support–especially Congress–do not yet appreciate the hazards involved in devoting support for science on the basis of immediate demands for particular results";[61] and the integrity of science, which was the primary concern of the report, remained threatened, in large part because of pressures on science to satisfy immediate social demands.[62] Furthermore, the committee supported the idea that scientific freedom was an "acquired right."[63] Stressing that "all citizens bear serious responsibilities toward the maintenance of the integrity of science," the report said: "Science needs to be protected from . . . constraints [or allowed to be free] by being permitted to develop and disseminate knowledge in keeping with its own procedures, so that the usefulness or desirability of this knowledge to society may be determined by that society."[64]

A third committee report, concerned with secrecy and its effects on the dissemination of scientific and technological knowledge, was issued in 1968.[65] This report acknowledged that science had progressed under secretive conditions in Nazi Germany and the Soviet Union, noted the argument that the U.S. defense-related research and development community was sufficiently large to permit wide peer discussion and circulation of security classified research,[66] and recognized the need–indeed, the value–of maintaining privacy in research involving human subjects.[67] It also contrasted basic science and technology in terms of secrecy practices. Pure research was conducted by self-selected and self-propelled scholars who were "relatively impervious to sanctions other than the approval or disapproval of . . . peers" exercising critical professional appraisal. To gain acceptance within the scientific community, these research results had to be publicly available. In the case of technology, certain professional canons, such as protection of industrial advantage, diplomatic discretion, client and patient confidentiality, or "loyalty to employing institutions where institutional aims are at stake" all mandated secrecy.[68] Nonetheless, the committee ultimately objected to "excessive and inappropriate uses of secrecy." These might be understood as disclosure restrictions that interfere with "the core scientific process" and penalize not only the scientific community, but also the whole society.[69]

The institutional successor to the Committee on Science in the Promotion of Human Welfare was the AAAS Committee on Scientific Freedom and Responsibility, a permanent advisory panel created in December 1970. It has functioned ever since as a watchdog against encroachments on scientific freedom and as a promoter of scientific responsibility.[70]

During the 1980s and early 1990s, the National Academy of Sciences created three temporary study panels to explore and report on scientific freedom issues. The first, called the Panel on Scientific Communication and National Security, was chaired by Dale R. Corson and functioned during 1982. The second, the Panel on the Impact of National Security Controls on International Technology Transfer, which was chaired by Lew Allen, Jr., did its work during 1986. The Academy was mandated by the Omnibus Trade and Competitiveness Act of 1988 to create a third study group.[71] The resulting Panel on the Future Design and Implementation of U.S. National Security Export Controls, chaired by Roland W. Schmitt, subsequently reported in early 1991. The findings and recommendations of these Academy committees are discussed in later chapters.

These study panel reports indicate that the organized scientific community has been thoughtfully aware of the conditions of freedom under which it operates and is sensitive to the influences, which may be somewhat subtle at times, affecting that environment. Perhaps *scientific freedom* has sometimes been idealized, but that is understandable in view of its fundamental importance. Furthermore, perhaps it is acceptable if the concept receives greater visibility, awareness of potential encroachments on it is heightened, and judgment regarding the degree of threat or resulting damage is not impaired by a lofty understanding of scientific freedom.

ACADEMIC FREEDOM

Closely allied with scientific freedom is another important value—academic freedom. As universities became the primary home of scholarship during the Middle Ages, and almost immediately the scholar's freedom to express new ideas led to tension because most universities—from medieval times well into the 19th century—were controlled by the church, innovative thinkers frequently found their institutions unsympathetic to ideas and knowledge that seemed to threaten ecclesiastical authority, the status of the church in society, or the established order. The struggle for academic freedom has been a long history of attempts to eliminate church or state control over the content of teaching and scholarship in colleges and universities.[72]

In many places, the so-called liberal-democratic revolution of the 18th and 19th centuries ended or largely abated such conflicts.[73] In the United States, the First Amendment to the Federal Constitution has been applied to questions of academic freedom.[74] Justice Felix Frankfurter, for example, outlined "four 'essential freedoms' of a university—to determine for itself on academic grounds who may teach, what may be taught, how it shall be taught, and who may be admitted to study."[75] This view only indirectly embraces research—as a possible source of subject matter for teaching, as a potential method of teaching, and as something university students might be involved in as study. Walter P. Metzger, a longtime student of academic freedom, maintained that academic freedom protections "extend to all in the class called faculty members, whether or not they do research," whereas scientific freedom protections "extend to all in the class called researchers, whether or not they are members of a faculty."[76] Nonetheless, the two values share some common ground and, indeed can be mutually supportive of each other. However, Metzger offered some important distinctions. For example, a theoretical explication of the rights and limits of scientific freedom, focusing on the relationship of science to society, needs to be devised. For academic freedom, such a formulation, concentrating on the relationship of university faculties to administrations, exists. Similarly, scientific freedom requires its own arrangements and procedures for detecting and adjudicating offenses that would not be confined to campus jurisdictions or concerned primarily with threats to employment.[77]

Until 1980–1981, it appears that the federal government made "little direct intrusion into research activities once begun," especially in universities.[78] This, however, was a pivotal period. Federal officials began to bring national security controls to bear on various facets of scientific communication on and off the campus. Both scientific and academic freedom were threatened. Accordingly, in late February 1981, the presidents of Cornell University, the Massachusetts Institute of Technology, the California Institute of Technology, the University of California, and Stanford University sent a jointly signed letter to the Secretaries of Commerce, State, and Defense, collectively, "to request clarification of the applicability of certain export restrictions to teaching and research activities conducted by American universities" and indicating that they were "deeply concerned about recent attempts to apply to universities the International Traffic in Arms Regulations (ITAR) and the Export Administration Regulations (EAR)." In concluding their letter, the university presidents expressed the "hope that after examining this issue carefully, you will clarify what has always been our understanding—namely, that the regulations are not intended to limit academic exchanges arising from unclassified research and teaching."[79]

The controversy, however, as made evident in subsequent chapters, continued and elevated before abating somewhat during the last years of the decade. It also revealed the truth of Metzger's observation: The rights and limits of scientific freedom, unlike academic freedom, still await developed theoretical expression.

SCIENTIFIC RESPONSIBILITY

Scientists in the United States are as legally and morally accountable as other citizens, but their professional abilities and knowledge also convey additional responsibilities to society, to their profession, and to the goal of scientific truth.

Maintaining the quality and integrity of the work of the scientific community is, of course, a basic responsibility of scientists.[80] This might be characterized as intellectual responsibility—utilizing critical examination and review to detect error and fraud.[81] When extended to technology—the quality of consumer products or the adequacy of a public structure, such as a bridge—its intertwining with moral and legal responsibilities is clearer. Such combined responsibilities have been formalized and institutionalized in licensing arrangements for certain science-based professions, such as medicine and engineering, and independent peer certification of some technology products. Professional codes of ethics reinforce such responsibilities and set standards regarding them.[82] The strong commitment of individuals to such responsibilities has contributed to the modern phenomenon of "whistle blowing," which has had serious consequences for particular scientific and technological undertakings, careers, and professions.[83]

The dimensions of scientific responsibility remain vague and confusing to both scientists and the public, however, due to differing viewpoints on the concept. The first of these, the *evaluative neutrality perspective*, considers scientific knowledge intrinsically valuable to the scientific community, but neither containing nor entailing moral rules. Utilizing scientific knowledge, a scientist or technologist may give technical advice about the most efficient or economical means to achieve certain ends, but such professional knowledge does not qualify either of them to assert that any one end is preferable to another. Indeed, because it is evaluatively neutral, though not without instrumental value, scientific knowledge may be used indifferently for any purpose, whether good or bad. The responsibility of the scientist, however, is not for the social or ethical consequences of such applications, but only to the scientific community and its work norms—maintaining the integrity of scientific knowledge, open communication, and producing new discoveries.[84]

Rejecting the contention that scientific knowledge is evaluatively neutral, the *scientistic perspective* asserts that, in addition to providing a desirable procedure for acquiring knowledge about the world, science presents substantive criteria of evaluation for all of the problems of life. Responsible for promoting the principles of scientific rationality and eliminating irrational beliefs, scientists must offset the negative social effects of science by increasing scientific knowledge contributing to a more "rational" and more "scientific" attitude on the part of policymakers contending with public issues.[85]

Finally, the *critical-interactionist perspective* regards scientific knowledge as being something more than factual findings, propositions, and theories of intrinsic value to only the scientific community. It is a human product, a constructed model based on truthful research, but still a limited view of the whole. Furthermore, there is recognition that a technological value may result. Thus, when producing scientific knowledge, scientists and scientific technologists are responsible not merely for the accuracy of their work, but they are also required to reflect critically on "the social system in which science is applied and upon the purpose for which science is used." In doing so, they engage in social dialogue and debate regarding the uses of scientific knowledge and their critical assessments derive from interaction with other persons and with nature.[86]

Although much more might be said about these three views of the responsibilities of scientists, their significance for this discussion lies in their diversity and conflict. In the United States, the time has long passed when scientists could afford to be naive or unconcerned about the social consequences of their contributions to scientific knowledge or the uses to which their work might be put. Various federal, state, and local laws and regulations set some standards in these regards. Certain aspects of nuclear energy research, for example, have long been regulated by the Atomic Energy Act. More recently, the field testing of genetically altered organisms has been subjected to federal control. Government support and regulation of science are discussed shortly.

Scientists and technologists are expected to be aware of their obligations as citizens—to obey the law and, when in disagreement with it, to seek adjustments through legitimate prescribed means. Adherence to the law alone, however, probably is not sufficient for scientific responsibility. There also appear to be expectations of faithfulness to higher moral and ethical standards. The identification, theoretical expression, and enforcement of such values are important considerations. Interest in these developments goes beyond individual scientific professional societies and is broadly based.

That scientific knowledge can have societal repercussions is widely recognized. Many people fear that science will accidentally or unwittingly unleash something uncontrollable that will result in great harm to the planet and its inhabitants. The "technological imperative"—that if something is technically possible, it should be done—prompts equal concern. Indeed, in many minds, scientific and technological development are no longer automatically identified with human progress.

However, there are also expectations that the scientific community will use its knowledge and skills to combat disease, hunger, and environmental pollution. In addition, American society assumes that the utilization of scientific knowledge for defense makes the country more secure.

Given that scientific knowledge can have both negative and positive societal repercussions, what is the responsibility of scientists? Proponents of a broad view of scientific responsibility—adherents of the critical-interactionist perspective—take both kinds of impact into consideration. In its discussion of the social responsibilities of scientists, the AAAS Committee on Science in the Promotion of Human Welfare acknowledged in 1960 that some uses of scientific knowledge could produce "secondary effects that cause unanticipated harm." Consequently, "application of new scientific advances," said the panel, "calls for social decisions which weigh the benefits against the disadvantages, and the public needs to have the facts relevant to such a decision."[87] Scientists bear a responsibility to not only critically evaluate the accuracy and societal repercussions of scientific and technological enterprise, but also assist the public, through education and expert advice, with making informed decisions regarding such matters. Furthermore, they are to use their professional skills and knowledge "to detect incipient problems before they become unnecessarily acute."[88] Thus, both the negative and positive dimensions of applications of scientific knowledge are addressed.

The development of scientific responsibility is not an academic exercise of little or no real consequence. The production and application of scientific knowledge can directly affect communities, societies, and the world. Scientists and technologists are being called upon by professional bodies, like the AAAS Committee on Science in the Promotion of Human Welfare, to not only be more aware of and sensitive to these repercussions, but to also develop, articulate, and practice accountability in these regards. Moreover, for scientists and technologists to be perceived as irresponsible or uncaring about their responsibilities may invite some loss of scientific freedom—more regulation, controls, and restriction—and restraint of scientific progress.

INTELLECTUAL PROPERTY PROTECTION

In a knowledge-based economy, science has become integrally related to human welfare and the structure and quality of everyday life. An important consideration in this regard is intellectual property rights, which refers to securing for a person, by law, the results of creative efforts of the mind. Preeminent among these protections are patenting and copyright. The "person" involved may be an individual scientist, a research team, or a corporate entity. Furthermore, some may still assert that a "scientist's claim to his intellectual 'property' is limited to that of recognition and esteem,"[89] but the contemporary record of experience strongly suggests otherwise, as the following recent assessment attests:

> Scientific knowledge has become, in effect, a commodity, vulnerable to commercial interests, public demands, and military controls. The disputes reflect recurrent changes in governmental policies toward research today in the direction of greater restrictions on the free flow of information. And they reflect structural changes in the profession of science—changes that themselves are a source of tension as the norms and expectations of the scientific community conflict with current economic and political realities[90]

Historically, as noted earlier, an important accommodation between open scientific and technological information and the protection of potentially profitable knowledge was reached in the late Middle Ages with the rise of intellectual property law. However, legal rights (or natural rights given recognition in declarative law), including intellectual property rights, are "moral justifications for limiting the freedom of another."[91] What kind of limitation is effected?

Copyright, first given statutory recognition in England in 1709, confers upon authors or creators of literary, dramatic, musical, and artistic works a monopoly to produce, reproduce, adapt, or perform their works for a limited time. The English law, known as the Statute of Anne, provided a model for 12 state statutes during the period when the United States was governed under the Articles of Confederation. Dissatisfaction with the Articles led to the inclusion of an important new provision, Section 8 of Article I in the Constitution, giving Congress power "To promote the Progress of Science and useful Arts, by securing for limited Times to Authors and Inventors the exclusive Right to their respective Writings and Discoveries." The first federal copyright statute was enacted in 1790.[92] In general, copyright gives scientists and technologists a degree of control over their written and recorded words and designs.[93]

Patenting is another arrangement for protecting intellectual property. A patent is a 17-year right of exclusive use given to an inventor in exchange for the disclosure of the invention so that it will be available for free public use when the patent period expires. Like trade secrets law, another protective arrangement discussed shortly, patenting grew out of concern for commercial ethics and seeks to provide incentives for innovation. The Republic of Venice is credited with enacting the first patent law in 1474. The inaugural American statute was produced by the Massachusetts Bay Colony in 1641.[94] The federal Congress received its authority to legislate patent law from Article I, Section 8 of the Constitution as well, and initially did so in 1790.[95]

All information in a patent application is held in confidence by the Patent and Trademark Office of the Department of Commerce until the patent actually issues. Approved patent applications may be inspected, but the information may not be utilized during the patent period except through licensing with the holder. Patent infringement or piracy may be pursued in court. However, this kind of litigation is highly complex, often prolonged, and usually expensive. For several reasons, patenting is less desirable than trade secret protection. Nevertheless, there are certain types of inventors who find these arrangements quite suitable for their purposes. Universities and small businesses, for example, may not be prepared to maintain a trade secret or otherwise may regard patenting a better way to realize profit through the licensing or sale of patent information. Also, in an area of rapid development, an inventor may consider 17-year exclusive use adequate, knowing that by the time the protection period lapses, innovation will provide patentable enhancements on the original invention or some other improvement making it obsolete.

Regarding the shift away from patenting, attorney Russell Stevenson, Jr., commented that, "while the empirical evidence is scanty, it appears that reliance on trade-secret protection is increasing, and in many cases even where inventions meet the standards of patentability."[96] He offers a number of reasons for this phenomenon. First, it is more expensive to acquire and defend a patent than to keep a new technological development protected by trade secret arrangements. Second, in addition to cost considerations, the law of trade secrets is thought to be more successful in safeguarding proprietary information. Third, because approved patent applications are available for public examination, there is a preference for the security that is afforded by trade secrecy. Finally, there are innovations that simply are not patentable.[97]

The concept of a trade secret, although probably very old, is not well defined or, perhaps better stated, its definition is not a commonly agreed upon understanding. Flexibility in interpretation is desirable so that new technologies and intellectual endeavors may be accommodated by the term and its underlying protective status. Probably the most widely accepted statement of what constitutes a trade secret is found in the *Restatement of Torts*, which, in part, says: "A trade secret may consist of any formula, pattern, device or compilation of information which is used in one's business, and which gives him an opportunity to obtain an advantage over competitors who do not know or use it."[98]

Given the somewhat loose nature of the trade secret concept, it should not be surprising that the law of trade secrets is rather pliable. It is largely judge-made, although some federal and state statutes provide remedies against the disclosure of trade secrets. Also, it is in the common law tradition, arising from usages and customs of immemorial antiquity and from the judgments and decisions of the courts recognizing, affirming, updating, and enforcing such past practices.

Historically, the law of trade secrets arose from a concern about commercial ethics. In brief, its purpose is largely to enforce standards of fair play in business conduct. There are, therefore, certain expectations that must be met in order for judicial protection or a statutory remedy to apply. First, it must be demonstrated that the information in question constitutes a trade secret—that it is used in one's business, provides a competitive advantage, and is exclusively held. Second, it must be shown that reasonable precautions were taken to prevent the information from becoming known by "proper" means. Third, the secrecy breach must be established: The information was obtained by "improper" means or in violation of a contractual or fiduciary obligation.

There are some other secrecy considerations worth mentioning. First, depending on the circumstances, a business may not seek recovery for the loss of a trade secret for fear that, in the course of a judicial proceeding, more details about the trade secret at issue might have to be disclosed or other confidential commercial information might have to be revealed. Protective orders, of course, may be a mitigating factor. Nevertheless, the situation serves to explain why trade secrets may be closely held by a select few within a business enterprise and also why litigation directly addressing loss of trade secrets is not always pursued, particularly when some other prosecutorial strategy of less commercial risk might be available.

Another consideration: Although the law of evidence affords trade secrets privilege from routine disclosure in litigation, it is not an absolute protection. In order to have a fair trial, judges have ordered the production of documents and testimony involving trade secret information. The privilege, therefore, is a qualified one.

Third, businesses have found contracts and agreements to be a useful and effective way to protect trade secrets. Certainly for practical reasons, physical safeguards, compartmentalization of knowledge, and the close holding of information by a select few are not always sufficient or adequate. A nondisclosure contract not only provides an element of flexibility in security arrangements, but also offers an alternative, and perhaps less risky, prosecutorial strategy for punishing trade secret losses. In brief, a business may find it more desirable to pursue a breach of contract lawsuit than to litigate directly on expropriation of a trade secret.

Finally, it should be somewhat apparent from this discussion that trade secrets are a peculiar kind of secret. In fact, the term is something of a misnomer. Trade secrets are often widely known. As a consequence of nondisclosure contracts, a great many employees in a particular business may be exposed to its trade secrets. Through licensing arrangements, a corporation may reveal trade secrets to another firm. Moreover, certainly more than one business has become aware that a competitor, through independent initiative, possesses and uses a trade secret that is not its exclusive knowledge.

That intellectual property protections present some problems for scientific freedom, particularly concerning openness in science, there is no question. However, there are some interesting prospects for accommodation. The AAAS Committee on Science in the Promotion of Human Welfare observed in 1969 that scholar-scientists entering the applied field become practitioners who "must learn the canons of their particular profession," which might require, among other limitations, "absolute secrecy where patents and industrial advantage are concerned."[99] To "learn" such expectations of secrecy does not necessarily mean that they are to be blindly accepted. (Challenge, of course, does not justify disobedience.) In deference to scientific freedom, the value of openness or, at least, greater degrees of openness to science (and industry) could be indicated and ways to allow entrepreneurially based scientists and technologists to discuss their work could be pursued. In this regard, scientific freedom constitutes an ideal, not an absolute.

For campus-based scientists and technologists, adherence to this ideal seems to be better reinforced. Calling openness "an abstraction," Robert M. Rosenzweig, president of the Association of American Universities, points out that the status of certain specific practices in this regard—the uninhibited and timely publication of research findings, the voluntary sharing of information about studies in progress, and the inclusion of students in all aspects of research—"as the accepted norm gives them substantial force."[100] Such openness became threatened when campus-based scientists and technologists increasingly engaged in the performance of research having a strong potential for yielding results of entrepreneurial value.[101] Accommodation was reached by relying on patenting as an appropriate intellectual property protection and contractual agreement, where a research sponsor desired sufficient opportunity to realize such a safeguard, to a slight delay in publishing results so that patenting could be effected. These arrangements, noted Rosenzweig, have become "so widespread as to have become routine." The practice of contractually requiring such a brief publication delay, he maintained, is not "intrinsically incompatible with the canon of openness," so long as its purpose is confined to providing an opportunity for patenting and not otherwise a means for determining whether or not results may be submitted, in whole or in part, for publication.[102]

Nonetheless, for science and technology, there are a great many unresolved conflicts concerning intellectual property protection, and still more continue to emerge. From her study of the controversy over intellectual property and other controls on scientific research today, sociologist Dorothy Nelkin concluded that, at least in the United States, the mediation of proprietary disputes involving scientific and technological information has been particularly encumbered by the tendency of claims to be made in terms of inalienable rights. An alternative and more viable approach, for her, is a negotiated contract—a product of compromise. "Thus," she wrote, "the concept of rights must be understood not as morally imperative but as socially defined; delimited by time, place, and situation; and correlated with responsibility." Key terms and relationships must be parlayed. In the case of scientific freedom, for example, traditional assumptions have ceased to be fully accepted. Other science policy decisions concerning regulation or controls will require mutual understanding and shared definitions of "legitimate" or "good" research that merits protection, as well as "national security" and the kinds of scientific or technological knowledge that fall within its ambit and for what end purpose. Finally, she urged closer and more consistent attention to questions of secrecy, which are central in the controversy over proprietary controls on scientific and technological information, and of no small consequence for both science and the public interest.[103]

FEDERAL REGULATION AND SUPPORT

Scientists and technologists, to reiterate a point made earlier, bear an obligation equal to that of any other citizen to obey the law. This duty consequently subjects scientists and technologists generally to government regulation. Furthermore, upon accepting some particular privilege or benefit from the government—such as employment or funds for the performance of research—they may be subject to additional regulatory authorities. In the United States, at the federal level, "the present governmentally imposed, legally enforceable constraints on research topics, procedures, and communications are relatively new," commented a 1986 assessment. "Most of the current regulatory schemes were developed alongside the post-World War II arrangements for Federal financing of research," it noted, "and were influenced by the political attitudes and assumptions governing those arrangements and how they were instituted."[104]

Indications are that, prior to World War II, many American scientists considered it improper, if not illegal, for the federal government to make research grants to private universities. This attitude is thought to be attributable, in part, to scientists' fear of losing autonomy in performing their work.[105] In 1945, Bell Labs president Frank B. Jewett, who was also the president of the National Academy of Sciences, expressed his opposition to the creation of the National Science Foundation in words reflecting this concern: "Every direct or indirect subvention by Government is not only coupled inevitably with bureaucratic types of control, but likewise with political control and with the urge to create pressure groups seeking to advance special interests."[106]

Expenditures for research and development in the United States during the decade prior to World War II reflect a very small federal contribution. That pattern changed dramatically and significantly with the war effort, as Table 1.1 indicates.[107]

Accompanying the new federal largess for research and development were various forms of regulation, including a number of security controls, which gained widespread acceptance among scientists in the special atmosphere of the war. Moreover, the experience gave federal officials a new awareness of the importance of science and technology for defending the nation and maintaining its security in the postwar world. From the end of the war through the 1960s, the federal defense establishment was the preeminent source of governmental research and development funds, as Table 1.2 indicates.[108]

Increased U.S. involvement in and escalation of the Vietnam War in the mid-1960s prompted criticism on campuses, by both students and

TABLE 1-1. Comparison of National Research and Development Expenditures, 1930–1945 (excluding atomic energy).

	Millions of dollars expended					*Pecentage of total expenditures*			
Year	*Total*	*Federal Govern-ment*	*Indus-try*	*Univer-sities*	*Other*[1]	*Federal Govern-ment*	*Indus-try*	*Univer-sities*	*Other*
1930	166	23	116	20	7	14	70	12	4
1932	191	39	120	25	7	20	63	13	4
1934	172	21	124	19	8	12	73	11	4
1936	218	33	152	25	8	15	70	11	4
1938	264	48	177	28	11	18	67	11	4
1940	345	67	234	31	13	19	68	9	4
1941–45 average	600	500	80	10	10	83	13	2	2

[1]State governments, private foundations and research institutes, including nonprofit industrial institutes.

faculty, of university performance of military-sponsored research. One result of this protest was divestiture by several major research universities—Columbia in 1967, Cornell in 1969, Stanford and the Massachusetts Institute of Technology (M.I.T.) in 1970, and Michigan in 1972—of military-supported classified research centers. Another consequence of the disenchantment with military support of basic research was the so-called Mansfield Amendment of 1969. Introduced by Senator J. William Fulbright, but vigorously championed by Senate Majority Leader Mike Mansfield, it was attached to the armed forces appropriation authorization and set the following limitation: "None of the funds authorized to be appropriated by this Act may be used to carry out any research project or study unless such a project or study has a direct or apparent relationship to a specific military function or operation."[109] In response, the Department of Defense reviewed 6,600 research projects and determined that approximately 4% of them did not meet the relevancy test of the Mansfield Amendment. These projects accounted for about $8.8 million of the Department's $223 million research and development budget. The following year, the amendment was modified to give the Pentagon somewhat broader latitude in its support of research.[110] "Despite this change in language," commented science historian Jeffrey K. Stine, "the Mansfield Amendment had a long-standing influence on how the Department of Defense justified its expenditures for scientific research by introducing both greater caution and uncertainty in the awarding of grants and contracts."[111] With the arrival of the Reagan Administration, however, military funding of research and development again assumed a commanding portion of total federal spending for these functions (see Table 1.2).[112]

Today, of course, funding conditions are but one of a variety of ways in which the federal government, both in its military and civilian capacities, regulates science and technology. Moreover, there are a number of reasons for the efforts at official control of research.[113] A 1986 report of the Office of Technology Assessment identified at least four justifications of federal regulation of research agendas.[114] The first justification is fulfilling political objectives, which the Mansfield Amendment well illustrates. The second is avoiding environmental damage, demonstrated by such statutes as the Atomic Energy Act, governing the possession, use, and disposal of radioactive substances by researchers,[115] and the Toxic Substances Control Act, setting controls for a multiplicity of chemicals at all stages of their development.[116] The third justification is promotion or avoidance of specific economic consequences. Examples include the Supreme Court's decision in the *Chakrabarty* case, upholding the patentability of an invention—a genetically engineered strain of bacteria—that otherwise met legal patent requirements even though it was alive;[117] amendments to the Patent Code allowing certain researchers making inventions with federal financial assistance to patent them;[118] and, more recently, the Orphan Drug Amendments to the Food, Drug, and Cosmetic Act that are designed to stimulate private research and development of pharmaceuticals for rare diseases.[119] The final justification for official controls on research is preserving moral values. A temporary 1974 statutory moratorium on research with living fetuses before or after abortion,[120] and 1978 amendments to the General Education Provisions Act preventing specific subject matter, teaching methods, psychological tests, or educational research from being used or conducted without parental consent are illustrative of this consideration.[121]

Similarly, it is easy to justify regulation over research procedures and protocols. One obvious goal is the protection of human health and safety. The Occupational Health and Safety Act, first enacted in 1970, protects research personnel who might be subject to unnecessary hazards in their work.[122] The special protections of the Atomic Energy Act and the Toxic Substances Control Act are quite pertinent in this regard, as are the regulations of certain federal agencies, such as the Departments of Defense,[123] Energy,[124] and Health and Human Services,[125] along with the Food and Drug Administration[126] and the Public Health Service.[127]

Another justification is to protect animals used in experimentation.[128] Protection of the environment constitutes a third reason for controls. Maintaining clean air[129] and clean water[130] are both important considerations, and, of late, as biotechnology has developed, efforts have been

TABLE 1–2. Summary of Outlays for the Conduct of Research and Development, 1949–1993

	In Millions of Dollars			*As Pecentage of Total Outlays*			*As Percentages of GGDP*			*In Billions of Constant (FY 1987) Dollars*		
Fiscal Year	*National Defense*	*Nondefense*	*Total*	*National Defense*	*Nondefense*	*Total*	*National Defense*	*Nondefense*	*Total*	*National Defense*	*Nondefense*	*Total*
1949	762	178	940	2.0	0.5	2.4	0.3	0.1	0.4	3.8	0.9	4.7
1950	772	282	1,054	1.8	0.7	2.5	0.3	0.1	0.4	3.9	1.4	5.3
1951	846	279	1,125	1.9	0.6	2.5	0.3	0.1	0.4	4.0	1.3	5.4
1952	1,204	291	1,495	1.8	0.4	2.2	0.4	0.1	0.4	5.6	1.4	7.0
1953	1,553	295	1,848	2.0	0.4	2.4	0.4	0.1	0.5	7.1	1.3	8.4
1954	1,551	297	1,848	2.2	0.4	2.6	0.4	0.1	0.5	7.0	1.3	8.4
1955	1,550	345	1,895	2.3	0.5	2.8	0.4	0.1	0.5	6.9	1.5	8.4
1956	1,898	428	2,326	2.7	0.6	3.3	0.5	0.1	0.6	8.1	1.8	9.9
1957	2,134	577	2,711	2.8	0.8	3.5	0.5	0.1	0.6	8.8	2.4	11.2
1958	2,459	729	3,188	3.0	0.9	3.9	0.5	0.2	0.7	9.9	2.9	12.8
1959	5,364	1,020	6,384	5.8	1.1	6.9	1.1	0.2	1.3	21.1	4.0	25.1
1960	5,937	1,385	7,322	6.4	1.5	7.9	1.2	0.3	1.4	22.8	5.3	28.1
1961	6,922	1,864	8,786	7.1	1.9	9.0	1.3	0.4	1.7	26.3	7.1	33.4
1962	7,090	2,747	9,837	6.6	2.6	9.2	1.3	0.5	1.8	26.5	10.3	36.7
1963	7,144	4,221	11,365	6.4	3.8	10.2	1.2	0.7	1.9	26.2	15.5	41.7
1964	7,865	5,931	13,796	6.6	5.0	11.6	1.3	0.9	2.2	28.5	21.5	49.9
1965	7,077	6,763	13,840	6.0	5.7	11.7	1.1	1.0	2.1	25.0	23.9	49.0
1966	7,097	7,821	14,918	5.3	5.8	11.1	1.0	1.1	2.0	24.4	26.9	51.3
1967	8,068	7,894	15,962	5.1	5.0	10.1	1.0	1.0	2.0	26.8	26.3	53.1
1968	8,544	7,628	16,172	4.8	4.3	9.1	1.0	0.9	1.9	27.4	24.4	51.8
1969	8,314	7,346	15,660	4.5	4.0	8.5	0.9	0.8	1.7	25.4	22.4	47.8

1970	8,021	7,132	15,153	4.1	3.6	7.7	0.8	0.7	1.5	23.2	20.6	43.8
1971	8,108	7,301	15,409	3.9	3.5	7.3	0.8	0.7	1.5	22.3	20.1	42.4
1972	8,837	7,466	16,303	3.8	3.2	7.1	0.8	0.7	1.4	23.1	19.5	42.6
1973	9,139	7,896	17,035	3.7	3.2	6.9	0.7	0.6	1.3	22.7	19.6	42.4
1974	9,406	8,028	17,434	3.5	3.0	6.5	0.7	0.6	1.2	21.7	18.5	40.3
1975	9,715	8,821	18,536	2.9	2.7	5.6	0.6	0.6	1.2	20.4	18.5	39.0
1976	9,819	10,169	19,988	2.6	2.7	5.4	0.6	0.6	1.2	19.2	19.8	39.0
TQ	2,455	2,745	5,200	2.6	2.9	5.4	0.6	0.6	1.2	4.6	5.2	9.8
1977	10,874	10,569	21,443	2.7	2.6	5.2	0.6	0.6	1.1	19.6	19.1	38.7
1978	12,077	12,455	24,532	2.6	2.7	5.3	0.6	0.6	1.1	20.3	20.9	41.2
1979	12,129	14,196	26,325	2.4	2.8	5.2	0.5	0.6	1.1	18.7	21.9	40.7
1980	14,643	15,592	30,235	2.5	2.6	5.1	0.6	0.6	1.1	20.7	22.1	42.8
1981	16,937	17,231	34,168	2.5	2.5	5.0	0.6	0.6	1.2	21.8	22.2	43.9
1982	19,809	14,850	34,660	2.7	2.0	4.6	0.6	0.5	1.1	23.7	17.8	41.5
1983	22,296	13,602	35,900	2.8	1.7	4.4	0.7	0.4	1.1	25.6	15.6	41.3
1984	25,765	15,221	40,986	3.0	1.8	4.8	0.7	0.4	1.1	28.4	16.8	45.1
1985	30,360	16,856	47,216	3.2	1.8	5.0	0.8	0.4	1.2	32.2	17.9	50.1
1986	35,656	16,485	52,141	3.6	1.7	5.3	0.8	0.4	1.2	36.7	17.0	53.7
1987	37,097	16,159	53,256	3.7	1.6	5.3	0.8	0.4	1.2	37.1	16.2	53.3
1988	38,032	18,068	56,100	3.6	1.7	5.3	0.8	0.4	1.2	36.7	17.4	54.1
1989[1]	40,366	20,394	60,760	3.5	1.8	5.3	0.8	0.4	1.2	37.3	18.8	56.1
1990[1]	41,078	22,732	63,810	3.3	1.8	5.1	0.8	0.4	1.2	36.5	20.2	56.6
1991[1]	41,669	24,296	65,965	3.1	1.8	5.0	0.7	0.4	1.2	35.6	20.7	56.3
1992 estimate[1]	44,031	26,824	70,855	3.0	1.8	4.8	0.8	0.5	1.2	36.5	22.2	58.7
1993 estimate[1]	45,610	28,847	74,457	3.0	1.9	4.9	0.7	0.5	1.2	36.6	23.1	59.7

[1]Includes off-budget Postal Service outlays of the following amounts (in millions of current dollars): 1989:64 1990:82 1991:116 1992:163 1993:206.

underway to perfect a regulatory framework concerning environmental releases of genetically engineered organisms.[131]

Finally, there are at least four justifications for federal regulation of the dissemination of scientific knowledge. One is to uphold scientific standards. The creation of formal arrangements facilitating the availability of certain scientific and technological knowledge—through, for example, the National Technical Information Service[132]—and its public accessibility—via, for example, the Freedom of Information Act[133]—have facilitated peer review and other evaluations of new scientific information.

Protection of a professional or economic interest is another basis for regulating the dissemination of scientific and technological knowledge. Federal patent law, as noted earlier, protects an inventor's intellectual property interest for a period of time, but permits details about a patented invention to be publicly known.[134]

A third justification for federal regulation of the dissemination of scientific and technological knowledge is to protect the health, privacy, and safety of individuals. Elaborate rules and safeguard arrangements have been developed regarding the informed consent and privacy of human subjects participating in research sponsored by federal agencies.[135]

Finally, federal regulation of the dissemination of scientific and technological knowledge has been justified in terms of protecting an interest of the United States concerning its external relations. This interest may be characterized generally as involving national defense, foreign relations, national security, economic security, or some combination of these elements. Among the oldest of the current pertinent authorities are statutory provisions, initially enacted in 1917, mandating that, in the event of an official determination that the disclosure of an invention by the granting of a patent "would be detrimental to the national security," the invention in question will be subjected to a secrecy order that may be renewed annually.[136] Consequently, the inventor loses the market value of his or her creation, probably has to surrender models of and documentation on the invention, and is subject to severe penalties for improperly disclosing to unauthorized persons any information covered by the secrecy order.

A statutory system of export controls has been operative since 1940. The Export Administration Act contains distinct national security, foreign policy, and short supply controls that are applicable to durable *goods* and to *technology*. The latter term is defined in the statute to mean:

> the information and knowhow (whether in tangible form, such as models, prototypes, drawings, sketches, diagrams, blueprints, or manuals, or in

intangible form, such as training or technical services) that can be used to design, produce, manufacture, utilize, or reconstruct goods, including computer software and technical data, but not the goods themselves.[137]

These and other authorities mandating national security limitation of the dissemination of scientific and technological information are discussed in more detail in a subsequent chapter.

OVERVIEW

At the outset of the 1980s, as détente vanished and U.S. relations with the Soviet Union became increasingly tense, the prevailing consensus for minimal federal national security restriction of scientific communication—limited and highly selective application of government controls, reinforced by voluntary and proprietary constraints—was transgressed. Consequently, controversy arose over whether or not actual conditions warranted or justified increased national security controls, and over the breadth of the new restrictions and their seemingly adverse impact on American achievement and progress in science and technology.

By way of introduction, this initial chapter examined some of the central values and concepts affected by increased national security regulation. Primary among these are scientific freedom and scientific responsibility. Although it appears that such important terms are not precisely defined, they are not without substance and foundation. Moreover, they are not absolutes, but instead are socially defined correlatives enjoying a high degree of traditional understanding and appreciation. Although the federal government, acting on behalf of American society, in effect sought to redefine both concepts, the scientific community responded, challenging this change in terms of its assumptions, execution, and effects. Assessment of this dispute continues in the next chapter with an examination of the conceptual basis for increased controls.

NOTES AND REFERENCES

[1]See Abraham Kaplan. *The Conduct of Inquiry*. San Francisco: Chandler Publishing Company, 1964; John G. Kemeny. *A Philosopher Looks At Science*. Princeton: D. Van Nostrand Company, 1959; Ernest Nagel. *The Structure of Science*. New York: Harcourt, Brace, and World, 1961.

[2]Thomas S. Kuhn. *The Structure of Scientific Revolutions*. Second edition. Chicago: University of Chicago Press, 1970, p. 1. For a more expansive view of science, see Peter J.

Denning. "The Science of Computing: A New Paradigm for Science." *American Scientist*, v. 75, November-December, 1987, pp. 572–573.

[3]U.S. Office of Scientific Research and Development. *Science—The Endless Frontier* by Vannevar Bush. Washington, DC: U.S. National Science Foundation, 1960; originally published 1945, pp. 18–19.

[4]Lord Rothschild. "Forty-Five Varieties of Research (and Development)." *Nature*, v. 239, October 13, 1972, pp. 373–378.

[5]See James Brian Quinn and Robert M. Cavanaugh. "Fundamental Research Can Be Planned." *Harvard Business Review*, v. 42, January-February, 1964, pp. 111–124.

[6]U.S. Congress. House. Committee on Science and Technology.Subcommittee on Science, Research, and Technology. *Scientific Communication and National Security*. Hearing, 98th Congress, 2d Session. Washington,DC: U.S. Govt. Print. Off., 1984, p. 154.

[7]*Ibid.*, p. 157.

[8]Although National Security Decision Directives are not published by the executive branch, the text of NSDD 189 is reproduced in full in an appendix to this book.

[9]Derek deSolla Price. "The Relations Between Science and Technology and Their Implications for Policy Formation" in Gabor Strasser and Eugene M. Simons, eds. *Science and Technology Policies*. Cambridge, MA: Ballinger Publishing Company, 1973, p. 150.

[10]U.S. Congress. House. Committee on Science and Technology. Subcommittee on Science, Research, and Technology. *Science Policy: A Working Glossary*. Fourth edition. Prepared by the Science Policy Research Division, Congressional Research Service, Library of Congress. Committee print, 95th Congress, 2d Session. Washington, DC: U.S. Govt. Print. Off., 1978, p. 99.

[11]J.W. Grove. "Science As Technology: Aspects of a Potent Myth." *Minerva*, v. 18, Summer, 1980, pp. 294, 311.

[12]U.S. Congress. House. Committee on Science and Technology. *Science Policy: A Working Glossary*, pp. 99–100.

[13]John T. Edsall. *Scientific Freedom and Responsibility*. Washington, DC: American Association for the Advancement of Science, 1975, p. x.

[14]*Ibid.*, p. xi.

[15]G.E.R. Lloyd. *Magic, Reason, and Experience*. Cambridge, UK: Cambridge University Press, 1979, p. 41.

[16]*Ibid.*, p. 228; Ernan McMullin. "Openness and Secrecy in Science: Some Notes on Early History." *Science, Technology, and Human Values*, v. 10, Spring, 1985, p. 15.

[17]William Eamon. "From the Secrets of Nature to Public Knowledge: The Origins of the Concept of Openness in Science." *Minerva*, v. 23, Autumn, 1985, p. 321.

[18]Sissela Bok. *Secrets*. New York: Pantheon Books, 1982, p. 154; also, generally, see J. Edward Mercer. *Alchemy: Its Science and Romance*. London: Macmillan Company, 1921; John Read. *Prelude to Chemistry; And An Outline of Alchemy*. New York: Macmillan Company, 1937; Sherwood Taylor.*The Alchemists*. Woodstock, NY: Beckman Publishers, 1977.

[19]Eamon, "From the Secrets of Nature to Public Knowledge," p. 322; McMullin, "Openness and Secrecy in Science," pp. 16–17.

[20]Eamon, "From the Secrets of Nature to Public Knowledge," pp. 333-334; David Hull. "Openness and Secrecy in Science: Their Origins and Limitations." *Science, Technology, and Human Values*, v. 10, Spring, 1985, p. 5; McMullin, "Openness and Secrecy in Science," p. 17.

[21]Eamon, "From the Secrets of Nature to Public Knowledge," p. 334.

[22]Hull, "Openness and Secrecy in Science," p. 5; McMullin, "Openness and Secrecy in Science," pp. 17-19.

[23]McMullin, "Openness and Secrecy in Science," pp. 15-16.

[24]Eamon, "From the Secrets of Nature to Public Knowledge," pp. 335–336; for the

original argument, see Edgar Zilsel. "The Genesis of the Concept of Scientific Progress." *Journal of the History of Ideas,* v. 6, June, 1945, pp. 325–349.

[25]See Paolo Rossi. *Philosophy, Technology and the Arts in the Early Modern Era.* New York: Harper and Row, 1970.

[26]Eamon, "From the Secrets of Nature to Public Knowledge," p. 337.

[27]*Ibid.,* pp. 328–329.

[28]Hull, "Openness and Secrecy in Science," p. 5; also see William Eamon and Francoise Paheau. "The Academia Segreta of Giolamo Ruscelli: A Sixteenth-Century Italian Scientific Society." *Isis,* v. 75, June, 1984, pp. 327–342.

[29]Hull, "Openness and Secrecy in Science," p. 6; McMullin, "Openness and Secrecy in Science," pp. 19–21.

[30]Hull, "Openness and Secrecy in Science," pp. 6-7; also see Roger Hahn. *The Anatomy of a Scientific Institution.* Berkeley: University of California Press, 1971; Martha Ornstein. The Role of Scientific Societies in the *Seventeenth Century.* Chicago: University of Chicago Press, 1913.

[31]Eamon, "From the Secrets of Nature to Public Knowledge," p. 343; also see Marie Boas Hall. "Oldenburg and the Art of Scientific Communication." *British Journal for the History of Science,* v. 2, December, 1965, pp. 277–290.

[32]Eamon, "From the Secrets of Nature to Public Knowledge," pp. 342–344; Hull, "Openness and Secrecy in Science," pp. 7–8; also, generally, see Thomas Birch. *History of the Royal Society of London.* New York: Johnson Reprint Corporation, 1968; originally published 1756.

[33]Hull, "Openness and Secrecy in Science," p. 8.

[34]Carl L. Becker. *Freedom and Responsibility in the American Way of Life.* New York: Alfred A. Knopf, 1947, p. 31.

[35]*Ibid.,* pp. 30–31.

[36]Edsall, *Scientific Freedom and Responsibility,* p. 5.

[37]Harold P. Green. "The Boundaries of Scientific Freedom." *Newsletter on Science, Technology, and Human Values,* June 1977, p. 17; also see Harold C. Relyea. "The Constitution and American Science." *Forum for Applied Research and Public Policy,* v. 2, Fall, 1987, pp. 106–116.

[38]See *Organization for a Better Austin* v. *O'Keefe,* 402 U.S. 415 (1970); *New York Times Company* v. *United States,* 403 U.S. 713 (1971).

[39]See *National Association for the Advancement of Colored People* v. *Button,* 371 U.S. 415 (1963).

[40]See *Shelton* v. *Tucker,* 364 U.S. 479 (1960); *United States* v. *Robel,* 389 U.S. 258 (1967).

[41]Green, "The Boundaries of Scientific Freedom," p. 18; also see Symposium. "Limits of Scientific Inquiry." *Daedalus,* v. 107, Spring, 1978.

[42]The committee was chaired by Maurice B. Visscher and its members included Philip Bard, Robert E. Cushman, Richard L. Meier, and James R. Newman, with Walter Gellhorn serving as a consultant. Generally, see Walter Gellhorn. *Security, Loyalty, and Science.* Ithaca, NY: Cornell University Press, 1950.

[43]Special Committee on Civil Liberties for Scientists. "Civil Liberties of Scientists." *Science,* v. 110, August 19, 1949, p. 177.

[44]*Ibid.,* p. 178.

[45]*Ibid.;* Cf. Gellhorn, *Security, Loyalty, and Science,* pp. 59–60.

[46]The committee was chaired by J.A. Stratton and its members included Robert F. Bacher, Laird Bell, Wallace O. Fenn, Robert F. Loeb, E. Bright Wilson, Jr., and Henry M. Wriston.

[47]Committee on Loyalty in Relation to Government Support of Unclassified Research. "Loyalty and Research." *Science,* v. 123, April 20, 1956, p. 661.

[48]*Ibid.,* p. 662.

[49]Chaired by Ward Pigman, the committee's members included Barry Commoner, Gabriel Lasker, Chauncey D. Leake, and Benjamin H. Williams.

[50]Interim Committee on the Social Aspects of Science. "Social Aspects of Science." *Science*, v. 125, January 25, 1957, pp. 144, 146.

[51]*Ibid.*, p. 145.

[52]*Ibid.*, p. 143.

[53]Committee on Science in the Promotion of Human Welfare. "Science and Human Welfare." *Science*, v. 132, July 8, 1960, pp. 72–73. At this time, the committee's members included Robert B. Brode, Harrison Brown, T.C. Byerly, Laurence K. Frank, H. Jack Geiger, Frank W. Notestein, Margaret Mead, and Dael Wolfle, with Barry Commoner as chairman.

[54]*Ibid.*, p. 69.

[55]See *Ibid.*, pp. 71–72.

[56]*Ibid.*, p. 70.

[57]*Ibid.*, pp. 70–71.

[58]*Ibid.*, pp. 71–72.

[59]At this time, the panel was still chaired by Commoner, but replacing Brown, Geiger, Notestein, and Wolfle were Ansley J. Coale, John T. Edsall, William T. Kabisch, and Walter Modell; Cf. note 53.

[60]Committee on Science in the Promotion of Human Welfare. "The Integrity of Science," *American Scientist*, v. 53, June, 1965, p. 193.

[61]*Ibid.*

[62]*Ibid.*, p. 197.

[63]See text accompanying note 36.

[64]Committee on Science in the Promotion of Human Welfare. "The Integrity of Science," pp. 196–197.

[65]The committee was now chaired by Margaret Mead and had an almost entirely new membership, including Robert McC. Adams, Richard H. Bolt, James D. Ebert, Hudson Hoagland, Martin Meyerson, Robert S. Morison, and a returned Dael Wolfle who joined Kabisch and Modell; Cf. note 59.

[66]Committee on Science in the Promotion of Human Welfare. "Secrecy and Dissemination in Science and Technology." *Science*, v. 163, February 21, 1969, p. 787.

[67]*Ibid.*, pp. 788–789.

[68]*Ibid.*, p. 788.

[69]*Ibid.*, p. 789.

[70]For a thoughtful, but unofficial, committee mission statement, see John T. Edsall's report cited at note 13.

[71]102 Stat. 1107, 1363.

[72]See Richard Hofstadter and Walter P. Metzger. *The Development of Academic Freedom in the United States*. New York: Columbia University Press, 1955.

[73]Becker, *Freedom and Responsibility in the American Way of Life*, p. 60.

[74]See Thomas I. Emerson. *The System of Freedom of Expression*. New York: Random House, 1970, pp. 593–626; Howard O. Hunter. "The Constitutional Status of Academic Freedom in the United States." *Minerva*, v. 19, Winter, 1981, pp. 519–568.

[75]*Sweezy* v. *New Hampshire*, 354 U.S. 234, 255 (1957).

[76]Walter P. Metzger. "Academic Freedom and Scientific Freedom." *Daedalus*, v. 107, Spring, 1978, p. 107.

[77]*Ibid.*, pp. 108–109.

[78]Hunter, "The Constitutional Status of Academic Freedom in the United States," p. 540.

[79]The full text of this letter appears in National Academy of Sciences. *Scientific Communication and National Security*. Washington, DC: National Academy Press, 1982, pp. 136–139.

[80]Edsall, *Scientific Freedom and Responsibility*, p. 8.

[81]See, generally, William Broad and Nicholas Wade. *Betrayers of the Truth: Fraud and Deceit in the Halls of Science*. New York: Simon and Schuster, 1983; Brock K. Kilbourne and Maria T. Kilbourne, eds. *The Dark Side of Science*. Proceedings of the 63rd Annual Meeting of the Pacific Division, American Association for the Advancement of Science. San Francisco: Pacific Division, American Association for the Advancement of Science, 1983; Alexander Kohn. *False Prophets: Fraud and Error in Science and Medicine*. New York: Basil Blackwell, 1986; National Academy of Sciences. *Responsible Science*. Washington, DC: National Academy Press, 1992; Albert Teich and Mark S. Frankel. *Good Science and Responsible Scientists*. Washington, DC: American Association for the Advancement of Science, 1992. Also see AAAS-ABA Conference of Lawyers and Scientists. Project on Scientific Fraud and Misconduct. *Report on Workshop Number One*. Washington, DC: American Association for the Advancement of Science, 1988; AAAS-ABA Conference of Lawyers and Scientists. Project on Scientific Fraud and Misconduct. *Report on Workshop Number Two*. Washington, DC: American Association for the Advancement of Science, 1989. For additional reading, see Marcel Chotkowski LaFollette, ed. *Ethical Misconduct inResearch Communication: An Annotated Bibliography*. Washington, DC: National Science Foundation, 1988.

[82]See Rosemary Chalk, Mark S. Frankel, and Sallie B. Chafer. *AAAS Professional Ethics Project: Professional Ethics Activities in the Scientific and Engineering Societies*. Washington, DC: American Association for the Advancement of Science, 1980. For additional reading, see Mark S. Frankel, ed. *Values and Ethics in Organization and Human Systems Development: An Annotated Bibliography*. Washington, DC: American Association for the Advancement of Science, 1987.

[83]See, for example, Edsall, *Scientific Freedom and Responsibility*, pp. 5, 35–40; Judith P. Swazey and Stephen R. Scher, eds. *Whistleblowing in Biomedical Research: Policies and Procedures for Responding to Reports of Misconduct*. Proceedings of a workshop sponsored by the President's Commission for the Study of Ethical Problems in Medicine and Biomedical and Behavioral Research, the Committee on Scientific Freedom and Responsibility of the American Association for the Advancement of Science, and Medicine in the Public Interest. Washington, DC: U.S. Govt. Print. Off., 1982. Of related interest are certain recent congressional proceedings: U.S. Congress. House. Committee on Energy and Commerce. Subcommittee on Oversight and Investigations. *Scientific Fraud*. Hearings, 101st Congress, 1st Session. Washington, DC: U.S. Govt. Print. Off., 1989; U.S. Congress. House. Committee on Government Operations. *Are Scientific Misconduct and Conflicts of Interest Hazardous to Our Health?* H. Rept. 101–688, 101st Congress, 2d Session. Washington, DC: U.S. Govt. Print. Off., 1990; U.S. Congress. House. Committee on Government Operations. Subcommittee on Human Resources and Intergovernmental Relations. *Federal Response to Misconduct in Science: Are Conflicts of Interest Hazardous to Our Health?* Hearing, 100th Congress, 2d Session. Washington, DC: U.S. Govt. Print. Off., 1988; U.S. Congress. House. Committee on Government Operations. Subcommittee on Human Resources and Intergovernmental Relations. *Is Science For Sale? Conflicts of Interest Vs. The Public Interest*. Hearing, 101st Congress, 1st Session. Washington, DC: U.S. Govt. Print. Off., 1989; U.S. Congress. House. Committee on Government Operations. Subcommittee on Human Resources and Intergovernmental Relations. *Scientific Fraud and Misconduct and the Federal Response*. Hearing, 100th Congress, 2d Session. Washington, DC: U.S. Govt. Print. Off., 1988; U.S.Congress., House. Committee on Science, Space, and Technology. Subcommittee on Investigations and Oversight. *Maintaining the Integrity of Scientific Research*. Committee print, 101st Congress, 2d Session. Washington, DC: U.S. Govt. Print. Off., 1990. Also see Associated Press. "Nobel Scientist Apologizes to Whistle-Blower." *Washington Post*, May 4, 1991, p. A6.

[84]See Henk Verhoog. "The Responsibilities of Scientists." *Minerva*, v. 19, Winter, 1981, p. 587.

[85]*Ibid.*, pp. 587–588.

[86]*Ibid.*, pp. 588–589.

[87]Committee on Science in the Promotion of Human Welfare. "Science and Human Welfare," p. 71.

[88]*Ibid.*

[89]Robert K. Merton. "Priorities in Scientific Discovery." In Robert K. Merton, ed. *The Sociology of Science.* Chicago: University of Chicago Press, 1973, p. 273.

[90]Dorothy Nelkin. *Science As Intellectual Property.* New York: Macmillan Publishing Company, 1984, p. 91. Also see Vivian Weil and John W. Snapper, eds. *Owning Scientific and Technical Information: Value and Ethical Issues.* New Brunswick, NJ: Rutgers University Press, 1989.

[91]See Herbert L. Hart. "Are There Any Natural Rights?" *Philosophical Review*, v. 64, April, 1955, pp. 175–191.

[92]1 Stat. 124.

[93]See Gerald Magavero. "The History and Background of American Copyright Law: An Overview." *International Journal of Law Librarians*, v. 6, July, 1978, pp. 151–158; Lyman R. Patterson. *Copyright in Historical Perspective.* Nashville, TN: Vanderbilt University Press, 1968.

[94]Russell B. Stevenson, Jr. *Corporations and Information: Secrecy, Access, and Disclosure.* Baltimore: Johns Hopkins University Press, 1980, p. 17.

[95]1 Stat. 109.

[96]Stevenson, *Corporations and Information*, p. 21.

[97]*Ibid.*, pp. 21–22.

[98]American Law Institute. *Restatement of Torts.* Vol. 4. St. Paul, MN: American Law Institute, 1939, p. 5 (section 757, comment b).

[99]Committee on Science in the Promotion of Human Welfare. "Secrecy and Dissemination in Science and Technology," p. 788.

[100]Robert M. Rosenzweig. "Research as Intellectual Property: Influences Within the University." *Science, Technology, and Human Values*, v. 10, Spring, 1985, pp. 41–42.

[101]See *Ibid.*, pp. 43–44.

[102]*Ibid.*, p. 48.

[103]Nelkin, *Science as Intellectual Property*, pp. 98-102.

[104]U.S. Office of Technology Assessment. *The Regulatory Environment for Science.* Washington, DC: U.S. Govt. Print. Off., 1986, p. 11.

[105]See Don K. Price. "Endless Frontier or Bureaucratic Morass?" In Gerald Holton and Robert S. Morison, eds. *The Limits of Scientific Inquiry.* New York: W.W. Norton and Company, 1979, pp. 75–92.

[106]J. Merton England. *A Patron for Pure Science.* Washington, DC: National Science Foundation, 1982, p. 35.

[107]From The President's Scientific Research Board. *Science and Public Policy.* Vol. 1. Washington, DC: U.S. Govt. Print. Off., 1947, p. 10.

[108]From U.S. Office of Management and Budget. *Budget of the United States Government: Fiscal Year 1993. Supplement, February 1992.* Washington, DC: U.S. Govt. Print. Off., 1992, p. 127.

[109]83 Stat. 204, 206.

[110]84 Stat. 905, 908; see U.S. Congress. House. Committee on Science and Technology. *A History of Science Policy in the United States, 1940-1985* by Jeffrey K. Stine. Committee print, 99th Congress, 2d Session. Washington, DC: U.S. Govt. Print. Off., 1986, pp. 61-63.

[111]U.S. Congress. House. Committee on Science and Technology. *A History of Science Policy in the United States, 1940-1985*, p. 63.

[112]See U.S. Congress. House. Committee on Science and Technology. *Science Support by the Department of Defense.* Committee print, 99th Congress, 2d Session. Washington, DC: U.S. Govt. Print. Off., 1986.

[113]See U.S. Office of Technology Assessment. *The Regulatory Environment for Science,* pp. 13-29; U.S. Congress. House. Committee on Science and Technology. *A History of Science Policy in the United States, 1940–1985,* pp. 25–77.

[114]See U.S. Office of Technology Assessment. *The Regulatory Environment for Science,* pp. 33–69.

[115]42 U.S.C. 2011–2296.

[116]15 U.S.C. 2601–2629.

[117]*Diamond* v. *Chakrabarty,* 447 U.S. 303 (1980); also, generally, see U.S. Office of Technology Assessment. *Commercial Biotechnology: An International Analysis.* Washington, DC: U.S. Govt. Print. Off., 1984, pp. 383–407.

[118]94 Stat. 3015, 3019.

[119]98 Stat. 2815, 2817–2820.

[120]88 Stat. 342, 353; also see U.S. Congress. Senate. Committee on Public Welfare. *Fetal Research, 1974.* Hearing, 93rd Congress, 2d Session. Washington, DC: U.S. Govt. Print. Off., 1974.

[121]92 Stat. 2143, 2355–2356.

[122]84 Stat. 1590.

[123]See 32 C.F.R. 219.

[124]See 10 C.F.R. 745.

[125]See 45 C.F.R. 46.

[126]21 C.F.R. 50.

[127]48 C.F.R. P.H.S. 380.

[128]See 7 U.S.C. 2131–2155.

[129]See 42 U.S.C. 7401 et. seq.

[130]See 33 U.S.C. 1251 et. seq.

[131]See U.S. Congress. House. Committee on Science and Technology. *The Environmental Implications of Genetic Engineering.* Committee print, 98th Congress, 2d Session. Washington, DC: U.S. Govt. Print. Off., 1984; U.S. Congress. House. Committee on Science and Technology. *Issues in the Federal Regulation of Biotechnology: From Research to Release.* Committee print, 99th Congress, 2d Session. Washington, DC: U.S. Govt. Print. Off., 1986.

[132]See 15 U.S.C. 1151–1157, 1525.

[133]See 5 U.S.C. 552; however, concerning some scientists' displeasure with such public access, see Nelkin. *Science as Intellectual Property,* pp. 31–45; Judith Lowitz Adler. "The Impact of FOIA on Scientific Research Grantees." *Columbia Journal of Law and Social Problems,* v. 17, Number 1, 1981, pp. 1–44.

[134]See 35 U.S.C. 100–376.

[135]See, for example, 10 C.F.R. 745.4–745.14 (Department of Energy); 21 C.F.R. 50.20–50.27 (Food and Drug Administration); 28 C.F.R. 22.21–22.28 (Department of Justice); 32 C.F.R. 219.4 (Department of Defense); 45 C.F.R. 46.103, 46.116–46.117 (Department of Health and Human Services).

[136]See 40 Stat. 394, now codified at 35 U.S.C. 181–188; 40 Stat. 411, 422, now codified at 50 U.S.C. App. 10(i).

[137]50 U.S.C. App. 2415(4); see 50 U.S.C. App. 2401–2419.

2

SCIENCE, TECHNOLOGY, AND NATIONAL SECURITY

Underlying most of the federal efforts during the 1980s at controlling traditional professional communication concerning science and technology was a major policy concept—*national security*. In modern government, no matter what the form, few terms have greater importance or influence. Moreover, during the past few decades, use of this expression in political parlance, policy, and law, particularly in the United States, has become more frequent and visible. Perhaps it is a phenomenon of fashion; perhaps there is a more considered motive.

For many, national security is probably understood intuitively. It is vaguely perceived as referring to the safety or survival of the country. More thoughtful explication can become difficult. That *national security* is not always readily susceptible to definition, interpretation, or comprehension may be both its strength and weakness as a policy concept. Its imprecise and general nature, on the one hand, can provide considerable flexibility for governmental action; its vagueness, on the other hand, may render the term senseless or subject to accidental or willful misinterpretation, resulting in ventures that are detrimental to the rights and privileges of the citizenry. Indeed, as one observer has commented, "in the name of national security there is much that can be done that cannot otherwise be done."[1]

A few years ago, a noted Canadian law professor began his study of the legal dimensions of national security for a commission investigating the Royal Canadian Mounted Police by confessing that he did not know what the term meant. Moreover, he did not think the government knew either, and suggested that *national security* "is one of those terms after which one should add the phrase, 'whatever that means'."[2] In the United States, however, it is doubtful that constitutionally guaranteed rights can tolerate such generosity of interpretation. Indeed, the term does not appear in the Constitution.

INTELLECTUAL FOUNDATION

The concept of national security has its intellectual foundation in the political philosophy of *raison d'etat*. Reason of state, according to the German historian Friedrich Meinecke, "tells the statesman what he must do to preserve the health and strength of the State."[3] In a more elaborate explanation, the American political scientist Carl Friedrich wrote that "reason of state is nothing but the doctrine that whatever is required to insure the survival of the state must be done by the individuals responsible for it, no matter how repugnant such an act may be to them in their private capacity as decent and moral men." Moreover, he said, it "is merely a particular form of the general proposition that means must be appropriate to the end, must, in other words, be rational in regard to the end, and that those means are the best which are most rational in being the most likely to succeed."[4]

On its face, national security means simply the security of the nation. However, what terms of reference has the concept embraced, and what has been its geographic scope? Looking back to the late 1970s, a somewhat limited definition of national security was formulated by the Department of Defense—"the condition provided by: *a.* a military or defense advantage over any foreign nation or group of nations, or *b.* a favorable foreign relations position, or *c.* a defense posture capable of successfully resisting hostile or destructive action from within or without, overt or covert."[5] By contrast, a more recent official Japanese report discussing "comprehensive national security" eschewed military terms for predominantly economic considerations.[6] Interestingly, the end of almost 50 years of Cold War rivalry between the United States and the Soviet Union prompted recent interest in an economic dimension to American understanding of national security and may, for some, constitute a new "competitiveness" basis for restricting traditional scientific communication.[7]

Other explanations of national security have reflected a variety of perspectives. For example, one political scientist, Gabriel Almond, viewed national security policy as "that part of a nation's foreign policy which is concerned with the allocation of resources for the production, deployment, and employment of what we might call the coercive facilities which a nation uses in pursuing its interests."[8] Another,

Samuel Huntington, saw it as anything done "to enhance the safety of the nation's social, economic, and political institutions against threats arising from other independent states."[9] General Maxwell Taylor, who, by his own admission, "spent most of his adult life in activities related to national security," explained the concept in terms of "national valuables," which include "current assets and national interests, as well as the sources of strength upon which our future as a nation depends." Among these, he felt some were "tangible and earthy; others are spiritual or intellectual," and "range widely from political assets such as the Bill of Rights, our political institutions and international friendships, to many economic assets which radiate worldwide from a highly productive domestic economy supported by rich natural resources." In his view, "the urgent need to protect valuables such as these . . . legitimizes and makes essential the role of national security."[10]

National security policy analysts Harold J. Clem and Stanley L. Falk saw a state's national security as "something to be defined within the context of its external environment and the international order." They contended that "in its search for security the state endeavors to relate with its external environment in a way conducive to the preservation of those political, social, and economic institutions which it values." However, because security "is a measure both of the absence of actual threats to acquired values, and of the absence of fear that such values are threatened," they maintain that "security is to be seen partly as a state of mind of a people and of its leaders (nor need it be the same for both groups)." Consequently, among the more highly influential factors, "national character, tradition, and prejudices play most important roles in shaping a nation's security outlook."[11]

In 1983, former White House national security adviser Richard V. Allen offered a characterization of national security, saying "it must include virtually every facet of international activity, including (but not limited to) foreign affairs, defense, intelligence, research and development policy, outer space, international economic and trade policy, monetary policy, and reaching deeply even into the domains of the Departments of Commerce and Agriculture." For this practitioner, national security "must reflect the presidential perspective, of which diplomacy is but a single component."[12]

Political scientist Richard Ullman advanced the contemporaneous view that:

> a threat to national security is an action or sequence of events that (1) threatens drastically and over a relatively brief span of time to degrade the quality of life for the inhabitants of a state, or (2) threatens significantly to narrow the range of policy choices available to the government of a state

or to private, nongovernmental entities (persons, groups, corporations) within the state.[13]

These Cold War viewpoints reflected very little consensus about the terms of reference of the national security concept, except that it included a military dimension. Such diversity of opinion certainly had implications for public policy. Furthermore, the situation did not become any clearer when addressing the question of geographic scope. For example, among the explanations of national security offered here, the definitions presented by the Defense Department and Almond appear to emphasize matters external to the United States. By contrast, Huntington, Clem and Falk, and Allen seem to see both overseas and domestic interests being involved. Huntington explicitly stated that internal security policy is a subfield of national security policy.[14] Agreeing with this perspective, another political scientist, the late Harold D. Lasswell, considered "how to maintain a proper balance between national security and individual freedom in a continuing crisis of national defense."[15] This significant issue was perceived long ago by James Madison who, in a letter to his friend Thomas Jefferson, observed: "Perhaps it is a universal truth that the loss of liberty at home is to be charged to provisions against danger, real or pretended, from abroad."[16]

POLICY HISTORY

Although the genesis of the term is uncertain, national security apparently had some currency 200 years ago. Around 1790, students at Yale University reportedly debated the question, "Does the national security depend on fostering domestic industries?"[17] Concurrently, John Jay was writing about "national safety" in Federalist Paper No. 3 "as it respects security for the preservation of peace and tranquility, as well as against dangers from *foreign arms and influence*, as from dangers and the *like kind* arising from domestic causes."[18]

Subsequently, in the course of discussing "the extent and variety of national exigencies, and the correspondent extent and variety of the means which may be necessary to satisfy them," Federalist Paper No. 23 gave recognition to the concepts of *emergency power*[19] and *executive prerogative*,[20] which would greatly affect *national security*.[21] The English philosopher John Locke had explored these matters in his *Second Treatise of Civil Government* (1690). The former he envisioned as inherent discretionary authority of the executive to respond, "for the good of the society," to a threatening or dangerous situation for which the legislative power had provided no relief and existing law did not grant necessary remedy.[22] The latter he described as the executive's "power to act according to discretion for the public good, without the prescription of law and sometimes even against it.[23] These views stood in the tradition *of raison d'etat*.[24] They were also accepted and used by American Presidents.[25] For example, Thomas Jefferson penned the following comment shortly after leaving the White House:

> The question you propose, whether circumstances do not sometimes occur, which make it a duty in officers of high trust, to assume authorities beyond the law, is easy of solution in principle, but sometimes embarrassing in practice. A strict observance of the written laws is doubtless *one* of the high duties of a good citizen, but it is not *the highest*. The laws of necessity, of self-preservation, of saving our country when in danger, are of high obligation. To lose our country by a scrupulous adherence to written law, would be to lose the law itself, with life, liberty, property and all those who are enjoying them with us; thus absurdly sacrificing the end to the means.[26]

Similarly, Abraham Lincoln, late in his presidency, offered this view:

> My oath to preserve the Constitution to the best of my ability, imposed upon me the duty of preserving, by every indispensable means, that government—that nation—of which that Constitution was the organic law. Was it possible to lose the nation, and yet preserve the constitution? By general law life *and* limb must be protected; yet often a limb must be amputated to save a life, but a life is never wisely given to save a limb. I felt that measures, otherwise unconstitutional, might become lawful, by becoming indispensable to the preservation of the constitution through the preservation of the nation. Right or wrong, I assumed this ground, and now avow it. I could not feel that, to the best of my ability, I had ever tried to preserve the constitution, if, to save slavery, or any minor matter, I should permit the wreck of government, country, and Constitution all together.[27]

In our own time, Richard Nixon contended that, for reasons of national security, he engaged in certain exercises of emergency power that were in the Jefferson-Lincoln tradition of Lockean prerogative.[28] Neither his means nor his ends in these situations were legally or socially acceptable. Nixon's actions seemed to be premised on a belief, steadily growing during prior administrations, that the United States had drifted into a condition of permanent emergency and that national security was determined largely, if not exclusively, by the executive. Reflecting on the consequences of such a view, Nixon's assistant, Egil

Krogh, chief of the White House "plumbers unit," commented when he was sentenced for perjuring himself about the burglary of the office of Daniel Ellsberg's psychiatrist:

> I see now . . . the effect that the term "national security" had on my judgment. The very words served to block critical analysis. It seemed at least presumptuous if not unpatriotic to inquire into just what the significance of national security was . . . . The discrediting of Dr. Ellsberg, which today strikes me as repulsive and an inconceivable national security goal, at the time would have appeared a means to diminish any influence he might have had in mobilizing opposition to the course of ending the Vietnam War that had been set by the President. Freedom of the President to pursue his planned course was the ultimate national security objective.[29]

A perusal of the federal statutes indicates that national security suddenly began to appear, with some frequency, as an undefined term in laws enacted around the time of United States involvement in World War I. However, its use in legislation adopted after 1919 then tapered off over the next decade. In the aftermath of World War II, it suddenly became prominent again in official pronouncements, presidential directives, statutes, and agency regulations.

During the years between the world wars, national security began to receive some thoughtful conceptual consideration.[30] In 1934, historian Charles A. Beard published *The Idea of National Interest*.[31] Assessing the conduct of American foreign policy, then primarily economic in scope and motivation, Beard felt it was unduly influenced by and representative of powerful subnational interests in the United States. As an alternative, he argued for a concept of national interest in foreign affairs that promoted the well-being of the country as a whole.

Whereas Beard urged a new national perspective in America's external relations, other scholars, such as Edward Mead Earle, Arnold Wolfers, and Nicholas Spykman, sought a better integration of U.S. military resources and foreign policy.[32] Their conceptual term for this amalgamation was *national security*.[33] From their efforts came the view of national security being "the ability of a nation to protect its internal values from external threats."[34] Subsequently, a whole field of study would come to be organized around the term.[35]

Reflecting this new scholarly thinking and the kind of related outlook advocated by Beard, the influential public affairs author and columnist Walter Lippmann wrote in 1943: "A nation has security when it does not have to sacrifice its legitimate interests to avoid war, and is able, if challenged, to maintain them by war."[36] Contending that, in the past, "there has been no serious and sustained conviction that American commitments and interests and ideals must be covered by our armaments, our strategic frontiers, and our alliances," he sought a foreign policy in the future that brought into balance, "with a comfortable surplus of power in reserve, the nation's commitments and the nation's power."[37] In Lippmann's view, this was national security.

Within government, opinions such as these arose during discussions of postwar military and foreign policy plans. For example, Secretary of the Navy James Forrestal, testifying before the Senate Committee on Military Affairs in 1945 on legislation concerning a new department and consolidation of the armed forces, stressed that he was using the term national security "consistently and continuously" rather than the more traditional national defense phrase.[38] He also explained that "it has been a fetish of mine that the question of national security is not merely a question of the Army and the Navy." He said:

> We have tried to take into account our whole potential for war, our mines, industry, manpower, research, and all the activities that go into normal civilian life. I do not think you can deal with this only by the War and Navy Departments. This has to be a truly global effort.
>
> That is the way we approached it, not from a limited area, but tried to bring in every element of our Government.[39]

He subsequently identified for the committee certain defects in "our security structure revealed by World War II":

> Our foreign policy and military policy were not always closely integrated.
>
> When the war started no adequate machinery for the mobilization of our material resources, productive capability, and manpower was in existence. There also was no adequate machinery for the adjustment of our civilian economic life to the requirements of total war.
>
> There were at times gaps in the translation of strategic plans into plans for materiel and personnel.
>
> There were weaknesses in the planning of materiel requirements and duplications in procurement both within and between the military departments.
>
> The coordination of military and other war budgets was not as thorough and detailed as would have been desirable.
>
> The first test of any plan for future national security is the extent to which it applies these lessons of the past. Any plan which neglects this requirement is misconceived. Any plan which fails to meet this requirement must be judged inadequate.[40]

In the course of his testimony, Forrestal returned to this consideration, telling the committee that "we have a list of eight requirements against which to measure any plan for national security":

> (1) Organized means for the integration of foreign and military policy.
>
> (2) Organizations in being for directing industrial mobilization and for reconciling industrial mobilization with natural resources.
>
> That means in particular that you don't create military demands beyond your capacity to fill them or that will do injury to other great and urgent demands. And that question of balance, in my view, is one of the most important considerations in war.
>
> (3) A more efficient organization for the translation of strategic requirements into requirements for materiel and personnel.
>
> (4) Provision for the coordination of military and other war budgets.
>
> (5) Adequate means for the elimination of waste and duplication in and between the military departments.
>
> (6) An efficient coordinated intelligence organization serving all Government departments and agencies.
>
> (7) An organizational means for fostering scientific research and development within the military departments and among civilian organizations.
>
> (8) Full opportunity for each branch of the military services to develop for its specialized task.[41]

Secretary Forrestal's particular plan for realizing his conception of national security was prepared, at his request and under his official sponsorship, by his former investment banking partner, Ferdinand Eberstadt, who had become a major figure in wartime mobilization efforts. In the transmittal letter accompanying his report, Eberstadt stated that "effective national security policy calls for active, intimate, and continuous relationships not alone between the military services themselves but also between the military services and many departments and agencies of Government."[42] He rejected, however, the prospect that "unification of the Army and Navy under a single head would improve our national security."[43] Stating that the "primary purpose of our Military Establishment is to serve and protect national security," this involved, in his estimate, not only protection of the United States, its territories, and possessions, but also support of "our international policies and fulfillment of our international commitments under the United Nations Charter, the Act of Chapultepec, the Monroe Doctrine, the occupation of two widely separated enemy countries—Germany and Japan—and otherwise."[44]

Eberstadt's report regarded the national security phrase as one of "changing content and scope,"[45] but explicitly maintained that "our international policy in the years ahead looks for national security through a United Nations organization for the maintenance of world peace."[46] In a few years, however, opinion would diverge on this aspect of attaining national security. Nonetheless, with the enactment of the National Security Act of 1947, most of the recommendations of the

Eberstadt report were realized, including the creation of such primary institutions as the National Security Council and the Central Intelligence Agency.[47]

The National Security Act also had symbolic significance. It marked the official entrenchment of the national security concept and gave formal foundation for the rise of the national security state within the American governmental environment. Preservation and perpetuation of the nation by any and all means became the government's principal mission; to carry it out, secrecy became essential.

The growing importance of national security as a concept of governance was reflected in some of the work of the Commission on Organization of the Executive Branch of the Government, chaired by former President Herbert Hoover. In early 1949, its Task Force on National Security Organization expressed two revealing views. First, "the problem of national security is not intermittent," said the Task Force, "but continuous, that it is not a secondary aspect of normal government activity but (as the authors of the Constitution clearly realized) a primary one, coloring and in many ways controlling nearly every other aspect of Government." Second, in view of this perspective, the Task Force urged a better understanding of the national security concept, saying:

> Americans have been accustomed in the past to think of national security too largely in purely military terms. Today it is obvious that valid national strategy must embrace all our natural resources of every kind—human, material, industrial, scientific, political, and spiritual. The armed forces are simply the cutting edge—a deterrent to hostile action in ordinary times but when used in war, a last and desperate resort. Military policy and preparation are vital, but they are only one part of national security policy as a whole, which, if it is to succeed, must continuously integrate political objectives, military plans, economic strength, and civilian organization into a comprehensive and carefully formulated national policy and purpose.[48]

"Military strength and efficiency is important," agreed the parent Hoover Commission, "but it is only one element of national security." The panel reaffirmed the need to subordinate the military to civilian authority, saying, "we must hold the military rigidly accountable to the President, the Congress, and the people . . . not only to safeguard our democratic traditions against militarism, but to insure that military policy shall be in close accord with national needs and national welfare."[49] With regard to the other elements of national security, however, the Commission was silent. Moreover, no definition of the concept was offered.

Nonetheless, national security apparently had adaptive value. In June 1949, for example, Sidney W. Souers, executive secretary of the National Security Council, described the concept as "a point of view rather than a distinct area of governmental responsibility."[50] The question was, whose point of view—that of the President, Congress, or the people—was to be accepted?

In 1952, Arnold Wolfers, who had been among those at Yale grappling with national security a decade or so earlier, noted the rise of a new attitude about the concept, reaching all the way back to Beard's thinking. *National security* assumed the character of "according more exclusive devotion to the narrower cause of the national self" as opposed to *national interest* which, it was feared, had become synonymous with the "interests of all mankind." Exponents of the new view of national security, said Wolfers, saw policymakers "sacrificing the less inclusive national community to the wider but in their opinion chimeric world community."[51] Thus, he argued, those who now invoke national security in policy deliberations often "are afraid their country underestimates the external dangers facing it or is being diverted into idealistic channels unmindful of these dangers." Consequently, there is the matter of perceiving enemies and the threat(s), both in degree and in kind, they pose to the United States. "Moreover," wrote Wolfers, "the symbol suggests protection through power and therefore figures more frequently in the speech of those who believe in reliance on national power than of those who place their confidence in model behavior, international cooperation, or the United Nations to carry their country safely through the tempests of international conflict."[52]

As these subtleties of interpretation were occurring, a number of statutes referring explicitly to national security were enacted through the years. A computer search found 390 uses of the term *national security* in the 1970 edition of the *United States Code*. Additional examples could be found in the uncodified laws. Slightly over 100 of these entries pertained to government institutions, such as the National Security Council or National Security Agency. Another score were section headings, cross-references, or references to the National Security Act. However, slightly more than 240 of the remaining citations were descriptions of a policy condition granting the President extensive discretionary authority—for example, "for reasons of national security," "in the interests of national security," or "for purposes of national security."[53] In none of these instances was national security defined, but the cumulative effect, nevertheless, was one of giving the executive overwhelming latitude to determine national security and, accordingly, what actions could be appropriately taken regarding it.

In pursuing interventionist military and foreign policies from the Cold

War era on, Presidents used this discretionary authority freely, and the power of the executive increased while congressional power was diminished. According to civil libertarian John Shattuck, what evolved was a

> national security framework within which postwar presidents have sought 'the freedom to pursue [their] planned course of action.' To make up for their lack of constitutional authority to act so freely, every president since Truman has relied on two doctrines to justify executive initiatives to protect national security: inherent presidential power and post-hoc congressional ratification.[54]

Shattuck concluded that the President's national security powers have become "powers to act in peacetime as if the country were at war. But since at least 1945 we have lived in a twilight zone in which the distinctions between war and peace are so blurred, and the instability of the world so constant that presidents have lacked any objective guideposts for the exercise of their national security powers."[55] Thus, the nation experienced a number of national security misadventures, not the least of which were the Bay of Pigs debacle, the Vietnam tragedy, events surrounding the Watergate incident, and perhaps the Iran-Contra affair.[56]

POLICY INTERPRETATION

The unresolved question persists: Who interprets the national security concept to determine public policy? Richard V. Allen has conceded a predominant role to the executive, saying this area "must reflect the presidential perspective." Few would probably disagree that defense officials, both civilian and military, would be centrally involved. Political scientist Harold D. Lasswell, however, long ago contended that, although the executive has "initial responsibility for the review and balancing of security policies," co-equal responsibility rested with Congress, the courts, and the citizenry or "public opinion." Moreover, he stressed the importance of the policymaking arrangements for realizing national security policy.

> Since all security policies entail risk, the public interest calls for the calculation of risk by a *procedure* that balances each policy against every policy and arrives at a judgment to which many minds have contributed. Only by developing proper procedures can public confidence be gained or vindicated in the long-run wisdom of the outcome. At any given moment, well-informed persons may disagree as to whether defense expenditures are too high or too low, or whether the over-all defense program is in

> balance or seriously unbalanced. The public must estimate the bias and competence of rival leaders. The public interest can be protected by the use of a procedure that takes conflicting views into account and subjects them to the discipline of debate and exposure to available knowledge.[57]

In late December 1973, Senator Charles Mathias delivered a major floor speech on national security and constitutional government. Those were tumultuous times. During the year, various congressional committees and a special prosecutor had been pursuing matters regarding and related to the Watergate break-in. By the end of 1973, the trail had led investigators to the White House doorstep, and preparations were underway in the House Committee on the Judiciary to develop articles of impeachment against the President.

Senator Mathias began his remarks by explaining that, in his view, "national security means the preservation of the well-being and safety of our Nation and all of our people," as well as "protecting against arbitrary rule by one man or cabal." Moreover, like Lasswell, he indicated that "what is necessary to assure domestic tranquility and to preserve national security, or indeed, even to determine what national security is, is the joint responsibility of the Congress, the President, the courts, and all others charged with a public trust." It is, he told his colleagues, "specifically the joint duty of the legislative and executive branches to decide, in accordance with their respective established procedures of due process, what is necessary to protect the integrity and well-being of the Nation."[58]

The Senator then noted that "'national security,' as it has been used in recent years, has in fact the opposite of its real meaning of peace or tranquility. The evocation of national security as it has been used in recent times has as its purpose, the dramatizing of tension," he said, "which in turn induces fear, fear of some foreign or domestic threat whose dimensions are unknown but whose magnitude is hinted at darkly as being beyond existing statutory or constitutional capabilities to meet that threat."[59] He reiterated that "whatever is done in the name of national security should be carried out only in conformity with the Constitution and the body of statute law and through properly designated instrumentalities." The consequences of doing otherwise were, to him, all too clear:

> A failure to use these agreed upon means of assuring domestic tranquility and an effective defense against potential foreign enemies as prescribed by law, and to create other secret mechanisms outside the law can only lead to a loss of national security and toward a national security state. Watergate serves as a severe warning of the dangers implicit in the use of arbitrary power.[60]

In the aftermath of Watergate, at least one legal scholar—the late Arthur S. Miller, who had offered the same advice to the Senate Select Committee on Presidential Campaign Activities—recommended that Congress "take appropriate action to clarify the meaning of 'national security,' establish its legitimate uses, and promulgate guidelines for its future application."[61] In fact, Congress devoted a considerable amount of energy and resources to controlling the exercise of national emergency powers—a kind of extraordinary, discretionary authority very similar to that wielded in the name of national security.[62] A special committee was formed in the Senate to study the situation. It discovered some 470 provisions of statutory law granting the executive extraordinary powers, many of which are ordinarily exercised by Congress.[63] Subsequently, in 1976, the National Emergencies Act was enacted to provide a degree of regulation, consistent with separation of powers arrangements, over the invocation and use of national emergency authorities.[64] This statute also eradicated a few provisions of emergency law thought to be particularly dangerous to civil liberties. No similar study or regulatory scheme, however, resulted regarding national security powers.

The meaning of the national security concept has been the explicit object of congressional attention in some particular cases during the past few years. In a 1973 report, for example, the House Committee on Government Operations criticized a presidential security classification order because its use of "national security" appeared to provide, due to its vagueness, an overly broad standard for official secrecy. It also conflicted with pertinent language in the Freedom of Information Act, and otherwise confused semantic and legal distinctions between it and "national defense."[65]

A 1980 report by the same committee took issue with the breadth of authority conveyed by the Invention Secrecy Act, saying: "The basis for issuance of a secrecy order—the opinion of an agency head that disclosure 'would be detrimental to the national security'—is subjective and absolute." In the committee's critical judgment, with "this broad standard, what might be deemed detrimental can vary widely with the times and the agency."[66] Accordingly, the report made the following recommendation: "Change the basis of issuance of a secrecy order from the opinion of an agency head that disclosure might or would be 'detrimental to the national security,' to a more demonstrable standard of damage to the national defense."[67]

Late in 1983, during House floor consideration of legislation revising and extending the Export Administration Act, an amendment was approved that defined the phrase, "detrimental to the national security of the United States," as used in the bill, to mean "likely to result in a significant reduction in the military capability of the United States or likely to result in a significant advance in the military capabilities of countries to which the goods or technology involved are controlled under this section."[68] Unfortunately, this attempt to better demarcate national security, which would have been a legislative precedent, ultimately failed. The 98th Congress did not complete action on the legislation revising the export law before its final adjournment. The provision interpreting the national security phrase did not appear in successor bills introduced during the next Congress or in the reform measure actually enacted.[69]

The federal courts, on occasion, also have sought to circumscribe the national security concept.[70] For example, a 1956 decision of the Supreme Court in the case of *Cole* v. *Young* concluded that national security was used in a particular statute "in a definite and limited sense and relates only to those activities which are directly concerned with the Nation's safety, as distinguished from general welfare . . . ."[71] In 1972, the Court ruled in the *Keith* case that national security does not override Fourth Amendment protections against warrantless wiretaps in domestic surveillances.[72] More recently, it determined that the claim of performing national security functions does not absolutely immunize the Attorney General from suit for damages arising out of allegedly unconstitutional conduct.[73]

NATIONAL SECURITY AND SCIENCE

During the course of his December 1973 floor remarks on the national security state, Senator Mathias commented that "secrecy cannot be equated with national security."[74] He knew, of course, that the executive had substantial discretionary authority to protect information and otherwise control its availability or dissemination. His remarks, however, seemed to suggest that wiser use of this authority could be made, particularly in view of the contemporary unfolding of events.

Many years earlier, as the allied armed forces were bringing World War II to an end, a similar observation had been made by the White House science adviser. "Basically," wrote Vannevar Bush in his report to the President on a program for postwar scientific research, "there is no reason to believe that scientists of other countries will not in time re-discover everything we now know." Consequently, he advised: "A sounder foundation for our national security rests in a broad dissemination of scientific knowledge upon which further advances can be more readily made than in a policy of restriction which would impede our further advances in the hope that our potential enemies will not catch up with us."[75]

Bush knew that scientific achievement and progress flourished in an atmosphere of accessible research results and open communication. Moreover, as the war effort had shown, scientific advances contributed both directly and indirectly to the security of the nation. Other government officials, however, subscribed to the viewpoint against which Senator Mathias later warned. Thus, as peace returned once again to the recent European and Asiatic-Pacific war zones, a struggle began in Washington over the realization of national security through a policy of protecting scientific discoveries and knowledge by concealment or through one of advancing science by supporting traditional practices of broadly disseminating findings, conclusions, and interpretations, as well as open criticism, improvement, and possible rejection.

Science and national security are not necessarily antagonistic to one another. There is a long tradition of federal financial support of scientific research, with controls normally limited to studies having clear military purpose. Conflicts over increased national security controls, however, did arise in the 1950s and again in the 1980s. Such changes in policy in the 1980s became apparent to scientists when federal officials began seeking to limit certain communications concerning campus scientific research, selected presentations or participants at scientific professional conferences, and the teaching of some scientific subject matter to certain foreign students. In general, the rationale for these actions was that some particular kind of information or knowledge, considered sensitive in a national security context, was being improperly imparted and, in order to prevent this conveyance, the government was exercising its regulatory powers. The principal authorities relied on in these efforts were export control laws and government research contract prerogatives. However, security classification powers, the Atomic Energy Act, and the Invention Secrecy Act were invoked as well.

The application of such controls derived, in large measure, from changing national security conditions as perceived by federal officials, who took the initiative through stricter enforcement and reinterpretation of existing authorities. Justifications for this approach were that past enforcement had been lax, confidential law enforcement and intelligence assessments had revealed aggressive new efforts by hostile nations to acquire our sensitive scientific and technological knowledge by any means, or United States relations with certain nations, such as those of the East bloc, had badly deteriorated and consequential security adjustments were required. Regardless of the merits of these arguments, critics of increased controls found themselves contending with a *fait accompli* and without a forum to mount a challenge immediately and conveniently.

Countermoves through Congress or the courts were partly frustrated by a lack of consensus within the scientific community about the significance of the new controls. The presidents of Cornell, MIT, CalTech, the University of California, and Stanford had sent a joint letter in late February 1981 to the Secretaries of Commerce, State, and Defense, collectively, "to request clarification of the applicability of certain export restrictions to teaching and research activities conducted by American universities." They pointed out the "significant practical difficulties" such restrictions posed for universities and, consequently, called for "a thorough assessment of the policy implications, the necessity and prospective effectiveness of the restrictions, the extent of disruption of the established role and operations of universities, and the serious legal and constitutional questions raised."[76]

Contrasting with this decisive response were the results of a March 1982 survey of a sample of 100 leading American scientists and engineers. In this study by the Committee on Scientific Freedom and Responsibility of the American Association for the Advancement of Science (AAAS), replies were received from "presidents or top corporate officials of IBM, Westinghouse, General Electric, General Motors, and Lockheed; from the presidents of Harvard University, University of Washington, University of Pennsylvania, Stanford University, and UCLA; from seven American Nobel Laureates; from the presidents of engineering and scientific societies affiliated with AAAS and from the chairpersons of several AAAS sections."[77] With regard to there being "a basic conflict between the principle of open scientific communication and national security," the survey found that "the respondents hold both concerns—open scientific communication and national security—as twin objectives, although they differ substantially in defining the best means of achieving these goals and the priority which they would assign to each objective."[78] Moreover, no larger comity was discernable regarding increased national security controls. "Although there is presently no unanimity or consensus around a single position," concluded the AAAS survey, "the various perspectives presented in the letters are still in a very early state of formulation. The balancing of interests which support national security objectives with concerns about the importance of maintaining open scientific communication procedures remains an ill-structured problem." Nonetheless, the survey found that "responses also indicate that the academic and corporate officials contacted through this poll feel strongly about the need to foster more dialogue with the officials who are developing our government's national security policies."[79]

Countermoves through Congress were also complicated somewhat by mixed congressional reaction to increased national security controls and jurisdictional problems. A subcommittee of the House Committee on

Government Operations had launched prescient oversight hearings in the spring of 1980 critically examining executive branch efforts at applying official secrecy to certain patent applications, cryptographic research, and the publication of information concerning the way an atomic bomb works—the so-called *Progressive* case.[80] However, the subject matter of these hearings largely did not include increased national security controls occurring after the Soviet invasion of Afghanistan in December 1979. Moreover, the panel lacked legislative jurisdiction to address the secrecy problems it identified.

In the spring of 1982, the Senate Permanent Subcommittee on Investigations began highly revealing oversight hearings on the inadequacies of federal agencies in controlling the unauthorized acquisition of U.S. high technology by the Soviet Union and its Warsaw Pact allies. These proceedings corroborated assertions about lax enforcement of export controls, including both current inadequacies and those of the recent past.[81]

Two subcommittees of the House Committee on Science and Technology jointly held a March 1982 oversight hearing on the impact of national security controls on science and technology. Although this forum offered a better opportunity for the scientific community to raise criticisms about increased controls, political leverage was lacking because the committee had no jurisdiction over the statutory authorities involved. Nonetheless, among the topics of discussion was the national security concept. Admiral Bobby R. Inman, Deputy Director of Central Intelligence, differentiated it from the *national interest*, but did not define or conceptualize either term.[82]

Asked to distinguish between national security and national interest, or to provide "a clearly defined, straightforward concept of the national interest," Robert Rosenzweig, vice president for public affairs at Stanford University, indicated to the subcommittees that he could not do so, but did think that it was "easier to talk about the national security than it is to talk about the national interest. National interest," he said, "like public interest, like a lot of other terms that are used frequently, tend to be used to describe the position of the speaker rather than some external reality that we can find out there. It's a difficult area," he concluded.[83]

The president of the National Academy of Sciences, Frank Press, who had served as science adviser to President Jimmy Carter, offered the following comment regarding national security:

> There is a narrow way to describe national security in terms of specific Soviet military systems that have integrated circuits, that were designed primarily by us, and that were obtained by them in some clandestine way or another. On the other hand, I believe we should take a broader view of

> our national security, which recognizes, for example, the role that university research plays . . . and the contribution that university research makes to our national security. The basic concept of the computer and the stored program computer communications theory, new kinds of materials, represent fields of research being developed on university campuses. The semiconductor industry itself recognizes that it is underinvested in basic research and it is now turning to universities to undertake that research with industry support. So these universities are a valuable national resource and here again we should balance the equation of information leakage on one side against the enormous contribution the universities make in their open and unrestricted research to our national security.[84]

Appearing before the subcommittees on behalf of the National Association of State Universities and Land-Grant Colleges, Robert Corell expressed both uncertainty and frustration regarding the national security concept, saying: "If one could be assured that national security information were to be classified only in those cases which clearly and unquestionably threaten our national security, and further, if there were a clear and common understanding of what we mean by national security, I don't think we would be here today."[85]

The comment by Dr. Press regarding industrial interest in and support of campus-based research served as something of an explanation for recently increased controls. In his testimony before the subcommittees, Lawrence J. Brady, Assistant Secretary of Commerce for Trade Administration, justified stricter enforcement of export restrictions saying:

> In the past, it has not been clearly understood that our technical data regulations apply equally to all segments of society—to industry, to individuals, and to academia. Moreover, as academic institutions have become increasingly involved in research for industrial applications, more technology becomes potentially subject to the regulations. Thus, as a consequence of academic involvement in research for industrial applications, we have become more aware of the impact on the national security of such transfers and have become more vigilant in enforcing the regulations.[86]

Testimony before the subcommittees also revealed that some senior executive branch officials were aware of the stultifying effects of increased national security controls. "We are caught in a dilemma," acknowledged George P. Milburn, Acting Deputy Under Secretary for Research and Development at the Defense Department. "If we vigorously attempt to regulate the flow of scientific information in the scientific community, it could jeopardize the strength and vitality of the very community we are seeking to revitalize for the sake of national defense. On the other hand," he noted, "if we abandon any attempt at regulation in the university context, it could seriously compromise and undercut our efforts to control the outflow of militarily-critical technology."[87]

Among those testifying on behalf of organizations and institutions representative of the American scientific community, opposition to increased national security controls was expressed in rather general terms. Dr. Press, for example, said:

> Most scientists and engineers in academia hold that open communications is absolutely essential for a creative research environment. In their view, restrictions on scientific communication sap the strength and vitality of the educational and research endeavors upon which the nation's present and future technology base is founded. Moreover, such restrictions are seen as inconsistent with a free society and, in some instances, with basic constitutional rights. Neither would they like to see us forced to emulate the Soviet practice of compartmentalizing and restricting access to knowledge.[88]

He also announced the creation of an Academy study panel—the Corson committee—which, with the cooperation and assistance of pertinent federal agencies, would examine the need for and effect of increased national security controls on the communication of unclassified, campus-based scientific research.[89]

Representatives of the scientific community once again discussed the national security concept before a congressional panel in November 1983. The occasion was provided by a subcommittee of the House Committee on the Judiciary holding oversight hearings on civil liberties and the national security state. Commenting on the need to attain "a balance satisfying reasonable national security needs and also perpetuating the rights and free expression and free ability of academics, of scientists and others, to pursue their work as they have traditionally in this country," subcommittee chairman Robert Kastenmeier said "we get such a diverse opinion as to what is national security, what is essential to national security, that I wonder whether we can achieve this balance."[90]

In response, C. Peter Magrath, president of the University of Minnesota, indicated that in his view, the definition of national security "has to come out of public debate and discussion and inquiry of the kind that your committee is undertaking in these hearings."[91]

Frank Press, who also participated in these proceedings, offered a comprehensive understanding of the concept.

> I would say there are many aspects in a definition of national security beyond military strength, although I don't think we should in any way

> diminish our ability to protect ourselves. I think we have to define it in terms of economic growth in the country, of the cultural life in the country, the quality of life of our citizens, and the example we portray to other nations in the world. Coming from the scientific area and the educational area, I happen to believe that these two fields have a great deal to do for our national security beyond our military strength. So I think we have to take a broad view of what constitutes national security.[92]

Another witness, Stephen Unger of the computer science department of Columbia University, dealt with national security in somewhat different terms, saying:

> The problem has frequently been falsely put in the form of a conflict between the personal rights and privileges of scientists and engineers and the national security. I submit that this is a false conflict in that there is no contradiction between these two interests.
>
> As has been stated by previous witnesses, the national security, to the extent that we define it in terms of military technology, is served by openness, not by secrecy. The same openness that enables scientists and engineers to exchange ideas freely with one another promotes the national security in the sense of enhancing the technological basis for the national security.
>
> I would further submit . . . that national security should not be narrowly defined in terms of military technology or military power. There is a good deal more to it than that.[93]

Unfortunately, such comments were not always heard by those to whom they were directed; moreover, when they were heard, they were not always heeded.

NATIONAL SECURITY POLICYMAKING

Suddenly confronted with increased national security controls on traditional scientific communication, some segments of the American scientific community responded through representatives who sought to ameliorate the situation. Their efforts, however, were frustrated by several dilemmas. There was a problem determining what was at issue. Neither the pertinent regulatory law nor the executive branch officials implementing stronger limitations offered a clear, thoughtful explanation of national security. Dialogue also proved to be elusive. To the extent that a view was discernable, national security seemed to be discussed almost exclusively in military terms. Moreover, there was little receptivity to a broader conception of national security—one regarding scientific enterprise, enhanced by traditionally open professional communication, as supportive of national security. Finally, those representing the scientific community found they lacked a policymaking forum or opportunity to mount a significant challenge.

These difficulties illustrate some of the larger problems attending national security policymaking. As earlier discussion has shown, national security is a vague concept, only loosely understood by public officials and devoid of statutory definition. Furthermore, it can be a dangerous concept—a *raison d'etat* for actions that may dearly cost the citizenry of a democracy many lives and liberties. The citizens of the American democracy, of course, have demonstrated, on various occasions, a willingness to make sacrifices, even supreme sacrifices, to maintain and preserve the security of their nation. There has long been an expectation, however, that national leaders will have exercised good judgment in determining the need for any such concessions, and that the reasons for their decisions may be readily known. In the aftermath of World War II, with the rise of the national security state in the United States, this confidence began to be betrayed. From the period of President Franklin D. Roosevelt's second term, beginning in 1937, through President Richard Nixon's first term, ending in 1972, there was an unprecedented expansion of presidential power regarding the external relations of the United States. As noted previously, a considerable amount of this growth occurred, in one way or another, in the name of national security. For various reasons, Congress acquiesced to or facilitated development, and, increasingly, a shroud of secrecy enveloped presidential decision making concerning national security. The result was that national security policy and actions were determined by an elite within the executive branch who were highly critical of any questioning of their judgment and also highly secretive about the reasons for their decisions.

If there was a single event that roused public and congressional reaction against this situation it was probably the conduct of the Vietnam War, including incursions by American armed forces into Cambodia and Laos, during the latter half of the 1960s.[94] Benchmark legislation reflective of a resurgent Congress actively seeking to reclaim co-equal status with the executive branch regarding the external relations of the United States came in 1973 with the War Powers Resolution, which was enacted over a presidential veto.[95] Contemporaneously, investigations of the Watergate affair revealed various improprieties undertaken for reasons of national security within the borders of the United States. Another congressional inquiry, begun in 1973, resulted in the National Emergencies Act of 1976, which restrained presidential use of statutorily delegated national emergency powers.[96] Moreover, in

1975, both Houses of Congress created temporary investigating committees to examine allegations of abuse and improper activities by U.S. intelligence agencies. Subsequently, both congressional chambers established permanent intelligence oversight committees, an annual authorization of intelligence community funding was instituted, and various intelligence reforms have been legislated from time to time. Indeed, Congress has become much more active and assertive in matters of foreign policy, war powers, and national security.[97]

Nonetheless, the executive branch and, particularly, the presidency, have not willingly relinquished their predominant status or readily accepted Congress as a co-equal partner in these policy matters. There is ample evidence of tension and conflict between the two branches in these regards during the Ford, Carter, Reagan, and Bush Administrations. Thus, citizens, such as scientists, engineers, or business persons, seeking adjustments or changes in national security policy can quickly find themselves embroiled in this interbranch rivalry. A more cynical view might characterize them as pawns in a high stakes policy game, provided the players choose to use them in their contest.

This situation warrants recall of Lasswell's important observation—"all security policies entail risk." Therefore, he suggested, "a *procedure* that balances each policy against every policy and arrives at a judgment to which many minds have contributed" is desirable for determining policy and action involving risk to the security of the nation.[98] Moreover, such a procedure should be presumptively open to citizen observation and participation. When secrecy is required, decisionmakers should still engage in the balancing process, and approve only policy and action, consistent with prevailing law, which, if known, could win the approval of the public (and, in the case of the executive branch, could also pass muster with Congress). "Only by developing proper procedures," counseled Lasswell, "can public confidence be gained or vindicated in the long-run wisdom of the outcome [of security policies and actions]."

The Constitution provides an obvious procedure in this regard—the legislative process of Article I, which provides for "determinations of policy that Congress can implement in only one way; bicameral passage followed by presentment to the President." Moreover, the Supreme Court has said constitutional history proffers "unmistakable expression of a determination that legislation by the national Congress be a step-by-step, deliberate and deliberative process."[99]

The President would also seemingly be well-served by a risk calculation procedure for determining security policy and action, one that, according to Lasswell's prescription, "takes conflicting views into account and subjects them to the discipline of debate and exposure to available knowledge." Although the Cabinet or, more particularly, the National Security Council might be an appropriate vehicle in this regard, neither has opened its deliberations to public scrutiny. This objective might be met, however, in some other way. For example, a few years ago, a House committee report offered the following pertinent recommendations concerning the determination of information security policy:

> that any future plans to revise security classification rules, whether by executive order or other executive action, be announced publicly;
>
> that any proposals to change security classification rules be circulated publicly for a period of at least sixty days and that public comments be accepted and considered by the President;
>
> that any proposals to revise security classification rules be provided to the Congress with sufficient time to permit interested congressional committees to consider the proposals, to hold hearings, and to prepare comments; [and]
>
> that the National Security Adviser or other official charged with policy responsibility for presidential security classification rules provide written findings detailing the problems that any proposed security classification rules changes are intended to solve, and written explanations to the Congress and to the public of the purpose and scope of any proposed changes. If the findings or explanations are classified, then a classified version should be provided to congressional committees and an unclassified version should be made public.[100]

Although these recommendations were proposed in the summer of 1982, they were never legislatively pursued and were otherwise ignored by the executive branch. The situation is a small reminder of the tensions and conflicts between the two branches regarding the determination of security policy and action. This rivalry, in no small part, can be traced to constitutional origins. Describing the distribution of political authority under the Constitution in a 1952 concurring opinion in the *Youngstown* case, Justice Robert H. Jackson wrote "there is a zone of twilight in which [the President] and Congress may have concurrent authority, or in which its distribution is uncertain."[101] It would appear that important national security, foreign affairs, and military powers lie in that twilight zone.

Nonetheless, without resorting to such drastic structural reforms as instituting a parliamentary form of government or establishing a Conseil d'Etat, some use of Lasswell's risk calculation procedure would seem to be viable for bridging interbranch disputes regarding security policy and action. For example, although a concurrent resolution adopted by both legislative chambers does not have the force of statutory law, such an instrument might be used to effect deliberation and discussion in Congress along the lines envisioned by Lasswell. Members might feel less partisanly constrained in commenting and voting on such a "sense of the Congress" resolution. Consequently, the President would have some indication of the extent to which his security policy or action, addressed by the concurrent resolution, was favored or opposed by legislators and why. Indeed, this approach recommends itself for being less confrontational than joint resolutions attempting to nullify or veto presidential policy or action. If the President ignored a concurrent resolution of disfavor and made no responsive adjustment, the adoption experience might provide congressional leaders a basis for gauging the prospects for taking binding action, such as exercising the power of the purse—a power assigned exclusively to Congress by the Constitution. "When the President takes measures incompatible with the express or implied will of Congress," counseled Justice Jackson, "his power is at its lowest ebb, for then he can rely only upon his own constitutional powers minus any constitutional powers of Congress over the matter."[102] Congress probably should not use its exclusive power of the purse to inhibit the President when he has independent constitutional authority to act. However, it can be utilized to halt initiatives outside of the President's constitutional authority *and* to effect better cooperation between the executive and legislative branches when Congress perceives that the President does not have exclusive authority.

Finally, for all of this particular discussion, there is an important Lasswellian footnote: "The public must estimate the bias and competence of rival leaders [regarding security initiatives]." Although such "leaders" may be representative of both different political parties and branches of government, the key term here is *must*: Recent past history sugests that security policy and action may well be doomed unless the citizenry comprehend and approve of them. The concurrent resolution example activates Lasswell's risk calculation procedure, enabling public scrutiny and possibly public participation in the decision making. It is this avenue of entry that representatives of the American scientific community so very much desired when atempting to ameliorate increased national security controls on traditional scientific communication.

NOTES AND REFERENCES

[1]Stuart Macdonald. *Technology and the Tyranny of Export Controls*. Houndmills: The Macmillan Press, 1990, p. 26.

[2]M.L. Friedland. *National Security: The Legal Dimensions*. Hull: Supply and Services Canada, 1979, p. 1.

[3]Friedrich Meinecke. *Machiavelliam: The Doctrine of Raison D'Etat and Its Place in Modern History*. New York: Frederick A. Praeger, 1965, p. 1.

[4]C.J. Freidrich. *Constitutional Reason of State: The Survival of the Constitutional Order.* Providence, RI: Brown University Press, 1957, pp. 4–5.

[5]Quoted from Maurice J. Montain. "The Continuing Complexities of Technology Transfer." *Government Executive*, v. 11, January, 1979, p. 42.

[6]See Robert W. Barnett. *Beyond War: Japan's Concept of Comprehensive National Security.* Washington, DC: Pergamon-Brassey's, 1984, pp. 8–10; Cf. Charles L. Schultz. "The Economic Content of National Security Policy." *Foreign Affairs*, v. 51, April, 1973, pp. 522–540.

[7]For examples of such interest, see B. R. Inman and Daniel F. Burton, Jr. "Technology and Competitiveness: The New Policy Frontier." *Foreign Affairs*, v. 69, Spring, 1990, pp. 116, 124: Theodore C. Sorensen."Rethinking National Security." *Foreign Affairs*, v. 69, Summer, 1990, pp. 7–11. Also see Theodore H. Moran. "International Economics and National Security." *Foreign Affairs*, v. 69, Winter, 1990–1991, pp. 74–90; Aaron L. Friedberg. "The Changing Relationship Between Economics and National Security." *Political Science Quarterly*, v. 106, Summer, 1991, pp. 265–276. Also, generally, see Graham Allison and Gregory F. Treverton, eds. *Rethinking America's Security: Beyond Cold War to New World Order*. New York: W.W. Norton and Company, 1992.

[8]Gabriel A. Almond. "Public Opinion and National Security Policy." *Public Opinion Quarterly*, v. 20, Summer, 1956, p. 371.

[9]Samuel P. Huntington. *The Soldier and the State: The Theory and Politics of Civil-Military Relations.* Cambridge, MA: The Belknap Press of Harvard University Press, 1957, p. 1.

[10]Maxwell D. Taylor. "The Legitimate Claims of National Security." Foreign Affairs, v. 52, April, 1974, p. 577.

[11]Harold J. Clem and Stanley L. Falk. *The Environment of National Security.* Washington, DC: National Defense University Press, 1977, p. 2.

[12]Richard V. Allen. "Foreign Policy and National Security: The White House Perspective." In Richard N. Holwill, ed. *Agenda '83: A Mandate for Leadership Report.* Washington, DC: The Heritage Foundation, 1983, p. 6.

[13]Richard H. Ullman. "Redefining Security." *International Security*, v. 8, Summer, 1983, p. 133.

[14]Huntington, *The Soldier and the State*, p. 1.

[15]See Harold D. Lasswell. *National Security and Individual Freedom.* New York: McGraw-Hill Book Company, 1950, pp. 1, 53–56.

[16]*Letters and Other Writings of James Madison.* Vol. 2. Philadelphia: J.B. Lippincott, 1865, pp. 140–141.

[17]W.W. Rostow. *How It All Began: Origins of the Modern Economy.* New York: McGraw-Hill Book Company, 1975, p. 191.

[18]See Benjamin Fletcher Wright, ed. *The Federalist.* Cambridge, MA: The Belknap Press of Harvard University Press, 1966, pp. 97–100.

[19]Concerning emergency power in the United States, see: Louis Koenig. *The Presidency and the Crisis.* New York: King's Crown Press, 1944; James G. Randall. *Constitutional Problems Under Lincoln.* Urbana: University of Illinois Press, 1951; Robert S. Rankin and Winfried R. Dallmayr. *Freedom and Emergency Powers in the Cold War.* New York: Appleton-Century-Crofts, 1964; Bennett Milton Rich. *The Presidents and Civil Disorder.* Washington, DC: The Brookings Institution, 1941; Clinton L. Rossiter. *Constitutional Dictatorship.* New York: Harcourt, Brace, and World, 1963; J. Malcolm Smith and Cornelius P. Cotter. *Powers of the President During Crisis.* Washington, DC: Public Affairs Press, 1960; U.S. Congress. Senate. Special Committee on National Emergencies and Delegated Emergency Powers. *A Brief History of Emergency Powers in the United States* by Harold C. Relyea. Committee print, 93rd Congress, 2d Session. Washington, DC: U.S. Govt. Print. Off., 1974.

[20]Concerning executive prerogative in the United States, see: Daniel P. Franklin. *Extraordinary Measures: The Executive Prerogative Powers in the United States.* Pittsburgh: University of Pittsburgh Press, 1991; Rankin M. Gibson. "The President's Inherent Emergency Powers." *Federal Bar Journal,* v. 12, October, 1951, pp. 107–151; James R. Hurtgen. "The Case for Presidential Prerogative." *University of Toledo Law Review,* v. 7, Fall, 1975, pp. 59–85; Arthur S. Miller. "Constitutional Law: Crisis Government Becomes the Norm." *Ohio State Law Journal,* v. 39, 1978, pp. 736–751; Note. "Honored in the Breech: Presidential Authority to Execute the Laws with Military Force." *Yale Law Journal,* v. 83, November, 1973, pp. 130–152; Note. "National Emergencies and the President's Inherent Powers." *Stanford Law Review,* v. 2, February, 1950, pp. 303–320; Donald L. Robinson. "The Routinization of Crisis Government." *Yale Review,* v. 63, December, 1973, pp. 161–174; John P. Roche. "Executive Power and Domestic Emergency: The Quest for Prerogative." *Western Political Quarterly,* v. 5, December, 1952, pp. 592–618; George Winterton. "The Concept of Extra-Constitutional Executive Power in Domestic Affairs." *Hastings Constitutional Law Quarterly,* v. 7, Fall, 1979, pp. 1–46.

[21]See Wright, *The Federalist,* pp. 199–203.

[22] John Locke. *Two Treatises of Government.* New York: Hafner Publishing Company, 1947, p. 203.

[23] *Ibid.,* p. 204.

[24]See text accompanying note 3.

[25]In his modern "stewardship" theory of the presidency, Theodore Roosevelt did not go so far as to embrace Locke's views, saying "the executive power was limited only by specific restrictions and prohibitions appearing in the Constitution and imposed by the Congress under its constitutional powers." See Theodore Roosevelt. *An Autobiography.* New York: Charles Scribner's Sons, 1926, pp. 388–389. A more restrictive interpretation of presidential authority was expressed by Roosevelt's Secretary of War and White House successor, William Howard Taft, who opposed the stewardship model. See William Howard Taft. *Our Chief Magistrate and His Powers.* New York: Columbia University Press, 1916, pp. 139–140; also see William Howard Taft. *The Presidency.* New York: Charles Scribner's Sons, 1916, pp. 125–130.

[26]H.A. Washington, ed. *The Writings of Thomas Jefferson.* Vol. 5. New York: H.W. Derby, 1861, p. 542.

[27]Roy P. Basler, ed. *The Collected Works of Abraham Lincoln.* Vol. 7. New Brunswick, NJ: Rutgers University Press, 1953, p. 281.

[28]See Richard Nixon. "Nixon—On Going beyond the Letter of the Law." *Washington Star,* June 5, 1977, pp. D-1, D-4.

[29]Quoted from John Shattuck. "National Security a Decade after Watergate." *Democracy,* v. 3, Winter, 1983, p. 57.

[30]See Morton R. Berkowitz and P.G. Bock, eds. *American National Security.* New York: Free Press, 1965, p. x; Morton Berkowitz and P.G. Bock. "National Security." In David L. Sills, ed. *International Encyclopedia of the Social Sciences.* Vol. 11. New York: The Macmillan Company and The Free Press, 1968, p. 40; P.G. Bock and Morton Berkowitz. "The Emerging Field of National Security." *World Politics,* v. 19, October, 1966, pp. 122–136; Daniel R. Yergen. *Shattered Peace: The Origins of the Cold War and the National Security State.* Boston: Houghton Mifflin Company, 1978, pp. 194–220.

[31]Charles A. Beard. *The Idea of National Interest.* New York: The Macmillan Company, 1934.

[32]Yergen, *Shattered Peace,* pp. 194–195, 450–451.

[33]*Ibid.,* pp. 194–195.

[34]See Morton Berkowitz and P.G. Bock, eds. *American National Security,* p. x; Morton Berkowitz and P.G. Bock. "National Security," p. 40; Arnold Wolfers. "'National Security' As An Ambiguous Symbol." *Political Science Quarterly,* v. 67, December, 1952, pp. 481–502.

[35]See P.G. Bock and Morton Berkowitz. The Emerging Field of National Security." *World Politics,* v. 19, October, 1966, pp. 122–136; also see Gene M. Lyons and Louis Morton. *Schools for Strategy: Education and Research in National Security Affairs*. New York: Frederick A. Praeger, 1965.

[36]Walter Lippmann. *U.S. Foreign Policy: Shield of the Republic*. Boston: Little, Brown and Company, 1943, p. 5.

[37]*Ibid.*, pp. 8–9.

[38]U.S. Congress. Senate. Committee on Military Affairs. *Department of Armed Forces, Department of Military Security*. Hearings, 79th Congress, 1st Session. Washington, DC: U.S. Govt. Print. Off., 1945, p. 99.

[39]*Ibid.*, pp. 8–9.

[40]*Ibid.*, pp. 577–578.

[41]*Ibid.*, pp. 578–579.

[42]See U.S. Congress. Senate. Committee on Naval Affairs. *Unification of the War and Navy Departments and Postwar Organization for National Security*. Committee print, 79th Congress, 1st Session. Washington, DC: U.S. Govt. Print. Off., 1945, pp. 1–2.

[43]*Ibid.*, p. 3.

[44]*Ibid.*, p. 15.

[45]*Ibid.*, p. 17.

[46]*Ibid.*, p. 18.

[47]61 Stat. 495; there was, however, a rebuff for Eberstadt: The new statute also provided for the administrative unification of the nation's armed forces and, ironically, Forrestal was named to be the first leader of this new establishment.

[48]U.S. Commission on Organization of the Executive Branch of the Government. *Task Force Report on National Security Organization*. Washington, DC: U.S. Govt. Print. Off., 1949, p. 28.

[49]U.S. Commission on Organization of the Executive Branch of the Government. *The National Security Organization.* Washington, DC: U.S. Govt. Print. Off., 1949, pp. 2–3.

[50]Sidney W. Souers. "Policy Formulation for National Security." *American Political Science Review,* v. 43, June, 1949, p. 535.

[51]Wolfers, "'National Security' As An Ambiguous Symbol," p. 482.

[52]*Ibid.*, p. 483.

[53]U.S. Congress. Senate. Committee on Governmental Affairs. *Emergency Powers of the District of Columbia.* Hearing, 96th Congress, 2d Session. Washington, DC: U.S. Govt. Print. Off., 1980, p. 18.

[54]Shattuck, "National Security a Decade after Watergate," pp. 61–62.

[55]*Ibid.*, p. 62; also see Saul Landau. The *Dangerous Doctrine: National Security and U.S. Foreign Policy*. Boulder, CO: Westview Press, 1988; Harold Hongju Koh. *The National Security Constitution: Sharing Power After the Iran-Contra Affair*. New Haven, CT: Yale University Press, 1990.

[56]To this list, some would also add the recent intervention of United States armed forces in the Persian Gulf: see, for example, Michael J. Glennon. "The Gulf War and the Constitution." *Foreign Affairs,* v. 70, Spring, 1991, pp. 84–101.

[57]Lasswell, *National Security and Individual Freedom,* p. 56.

[58]*Congressional Record,* v. 119, December 20, 1973, p. 42672.

[59]*Ibid.*

[60]*Ibid.*

[61]Arthur Selwyn Miller. "Implications of Watergate: Some Proposals for Cutting the Presidency down to Size." *Hastings Constitutional Law Quarterly,* v. 2, Winter, 1975, pp. 52–54.

[62]See U.S. Congress. Senate. Special Committee on the Termination of the National Emergency. *National Emergency*. Hearings, 93rd Congress, 1st Session. Washington, DC:

U.S. Govt. Print. Off., 1973; also see U.S. Congress. Senate. Special Committee on National Emergencies and Delegated Emergency Powers. *National Emergencies and Delegated Emergency Powers*. S. Rept. 94-922, 94th Congress, 2d Session. Washington, DC: U.S. Govt. Print. Off., 1976.

[63]See U.S. Congress. Senate. Special Committee on the Termination of the National Emergency. *Emergency Powers Statutes*. S. Rept. 93-549, 93rd Congress, 1st Session. Washington, DC: U.S. Govt. Print. Off., 1973.

[64]90 Stat. 1255; see U.S. Congress. Senate. Committee on Government Operations [and] Special Committee on National Emergencies and Delegated Emergency Powers. *The National Emergencies Act (Public Law 94-412). Source Book: Legislative History, Texts, and Other Documents*. Committee print, 94th Congress, 2d Session. Washington, DC: U.S. Govt. Print. Off., 1976. The text of the National Emergencies Act may be found at 50 U.S.C. 1601-1651.

[65]See U.S. Congress. House. Committee on Government Operations. *Executive Classification of Information—Security Classification Problems Involving Exemption (b)(1) of the Freedom of Information Act (5 U.S.C. 552)*. H. Rept. 93-221, 93rd Congress, 1st Session. Washington, DC: U.S. Govt. Print. Off., 1973, pp. 61-62.

[66]See U.S. Congress. House. Committee on Government Operations. *The Government's Classification of Private Ideas*. H. Rept. 96-1540, 96th Congress, 2d Session. Washington, DC: U.S. Govt. Print. Off., 1980, p. 3.

[67]*Ibid.*, p. 32.

[68]See *Congressional Record*, v. 129, September 29, 1983, p. 26485.

[69]See 99 Stat. 120.

[70]There are, however, also countervailing trends: See, for example, Jordan J. Paust. "Is the President Bound by the Supreme Law of the Land?—Foreign Affairs and National Security Reexamined." *Hastings Constitutional Law Quarterly*, v. 9, Summer, 1982, pp. 719-772.

[71]351 U.S. 536, 542-543 (1955).

[72]*United States v. United States District Court for the Eastern District of Michigan (Plamondon)*, 407 U.S. 297 (1972).

[73]*Mitchell* v. *Forsyth*, 472 U.S. 511 (1985).

[74]*Congressional Record*, v. 119, December 20, 1973, p. 42673.

[75]U.S. Office of Scientific Research and Development. *Science—The Endless Frontier* by Vannevar Bush. Washington, DC: National Science Foundation, 1980; originally published 1945, p. 190; Cf. *Ibid.*, p. 29.

[76]The full text of the letter appears in National Academy of Sciences. *Scientific Communication and National Security*. Washington, DC: National Academy Press, 1982, pp. 136-139.

[77]American Association for the Advancement of Science. "National Security and Scientific Communication: A Summary of Responses Received in Reply to a Letter from the AAAS Committee on Scientific Freedom and Responsibility." Washington, DC: American Association for the Advancement of Science, June, 1982, p. 1.

[78]*Ibid.*, p. 2.

[79]*Ibid.*, p. 8.

[80]See U.S. Congress. House. Committee on Government Operations. *The Government's Classification of Private Ideas*. Hearings, 96th Congress, 2d Session. Washington, DC: U.S. Govt. Print. Off., 1981; U.S. Congress. House. Committee on Government Operations. *The Government's Classification of Private Ideas*. H. Rept. 96-1540, 96th Congress, 2d Session. Washington, DC: U.S. Govt. Print. Off., 1980.

[81]See U.S. Congress. Senate. Committee on Governmental Affairs. *Transfer of United States High Technology to the Soviet Union and Soviet Bloc Nations*. Hearings, 97th Congress, 2d Session. Washington, DC: U.S. Govt. Print. Off.,1982; U.S. Congress. Senate.

Committee on Governmental Affairs. *Transfer of United States High Technology to the Soviet Union and Soviet Bloc Nations*. S. Rept. 97-664, 97th Congress, 2d Session. Washington, DC: U.S. Govt. Print. Off., 1982.

[82]U.S. Congress. House. Committee on Science and Technology. *Impact of National Security Considerations on Science and Technology*. Hearing, 97th Congress, 2d Session. Washington, DC: U.S. Govt. Print. Off., 1982, p. 8.

[83]*Ibid.*, p. 103.

[84]*Ibid.*, pp. 103–104.

[85]*Ibid.*, pp. 112–113, 117.

[86]*Ibid.*, pp. 9–10, 13–14.

[87]*Ibid.*, pp. 17, 25–26.

[88]*Ibid.*, p. 63, 70–71.

[89]See *Ibid.*, pp. 64–65, 80–84.

[90]U.S. Congress. House. Committee on the Judiciary. *1984: Civil Liberties and the National Security State*. Hearings, 98th Congress, 1st and 2d Sessions. Washington, DC: U.S. Govt. Print. Off., 1984, p. 87.

[91]*Ibid.*

[92]*Ibid.*, pp. 87–88.

[93]*Ibid.*, p. 118.

[94]See U.S. Congress. Senate. Committee on Foreign Relations. *The U.S. Government and the Vietnam War*. Committee prints, 98th Congress, 2d Session [and] 100th Congress, 2d Session. Washington, DC: U.S. Govt. Print. Off., 1984, 1985, 1988; U. S. Congress. House. Committee on Foreign Affairs. *The War Powers Resolution*. Committee print, 97th Congress, 2d Session. Washington, DC: U.S. Govt. Print. Off., 1982.

[95]87 Stat. 555.

[96]90 Stat. 1255.

[97]See, generally, Barry M. Blechman. *The Politics of National Security*. New York: Oxford University Press, 1990. For a recent dissenting view, however, see Michael J. Glennon. "The Gulf War and the Constitution." *Foreign Affairs*, v. 70, Spring, 1991, pp. 84–101.

[98]Lasswell, *National Security and Individual Freedom*, p. 56.

[99]*Immigration and Naturalization Service* v. *Chadha*, 462 U.S. 919, 954–955, 959 (1983).

[100]U.S. Congress. House. Committee on Government Operations. *Security Classification Policy and Executive Order 12356*. H. Rept. 97-731, 97th Congress, 2d Session. Washington, DC: U.S. Govt. Print. Off., 1982, p. 35.

[101]*Youngstown Sheet and Tube Company* v. *Sawyer*, 343 U.S. 579, 637 (1952).

[102]*Ibid.*

3

THE EMERGENCE AND APPLICATION OF CONTROLS

Efforts by the federal government during the 1980s to exert greater control over scientific communication relied on various national security authorities that accumulated over many years. The oldest of them trace their origins to the World War I era, when the military value of science and technology began to be better appreciated. Generally, these authorities are of four types: statutory law, presidential directives, administrative regulations, and contract provisions. From time to time, judicial pronouncements have reinforced or qualified their effect. Such rulings are an important reminder that these authorities are not immune to legal challenge, although attending criminal and civil penalties probably make few eager to disobey the law in order to bring a test case. However, they must pass constitutional muster, must be properly constructed, and, in their implementation, must be correctly interpreted.

The evolution, status, and requirements of these regulatory authorities are the subject matter of this chapter. So, too, are the countervailing legal protections pertaining to scientific communication in the United States. The chapter begins, however, with a brief overview of the World War I experience, when many within the scientific and technological community offered their talents for the defense of the country. Some important organizational mechanisms provided federal officials valuable opportunities to learn of the actual and potential military value of science and technology. The war experience also resulted in a legacy of government controls.

The next world conflict convinced the nation's leaders of the value of scientific and technological enterprise for defending the country. The arsenal of controls also increased, and continued to expand with the onset of the Cold War. Old temporary regulatory authorities were given permanent status to protect "national security." Simultaneously, there appeared to be no limit to the number and kinds of contributions of science and technology that were perceived by federal officials to be vital to the national security of the United States and, therefore, in need of control to limit their availability. Furthermore, a new regulatory capability was developing as federal funding for research and development—largely through the military departments—began its steady growth.

Scientific communication remained stringently restricted during the early 1950s. Relief from these security controls occurred as Cold War anticommunism gave way to international trade and a foreign policy of greater cooperation with the Soviet Union and its Warsaw Pact allies during the next decade. The legacy of control authority, however, remained available. As U.S. relations with the East bloc deteriorated during the late 1970s, culminating with the end of detente on the occasion of the Soviet invasion of Afghanistan, followed by a new, strongly anticommunist administration in Washington, the embers of control authority were fanned to flame.

THE WORLD WAR I EXPERIENCE

When writing his introduction for a 1920 compilation of assessments of the development of science in the United States during the period of World War I, George Ellery Hale commented "that so long as the security of the world is menaced by unscrupulous military powers, research methods must be effectively utilized in perfecting the means of national defense."[1] Director of the Mount Wilson Observatory and an active proponent of scientific enterprises, Hale had initiated the idea in July 1915 that the National Academy of Sciences, which included the nation's most preeminent scientists among its members, should offer its services to the President in the event the country went to war.[2] His statement should not suggest, however, that, heretofore, American science and technology had not made any significant contribution to the country's military capability.[3] "But," observed science historian Jeffrey K. Stine, "more than any other conflict before it, World War I punctuated the fundamental application of modern science and engineering to warfare. It became a war influenced heavily by the work of chemists, physicists, and electrical engineers.[4]

When war broke out in Europe in August 1914, the United States immediately proclaimed its neutrality. It soon became evident, however, that maintaining this position would be difficult, particularly with regard to American shipping. Within a few days after hostilities ensued, Great Britain effected a naval blockade of Germany. By the end of the year, all neutral vessels bound for or coming from Germany, Holland, and the Scandinavian countries were subject to Royal Navy interception and their cargoes were liable to seizure and confiscation.

In February 1915, Germany declared the waters around the British Isles a war zone. Enemy merchant ships found in this area would be destroyed on sight without provision for the safety of passengers and crew; neutral vessels entering this area did so at their own risk. That the Imperial Navy was using submarines in pursuit of the war at sea soon became well known. For many in the United States, however, the submarine menace took on dramatic proportions in early May when the British transatlantic steamer *Lusitania* was torpedoed without warning off the coast of Ireland and sunk with the loss of almost 1,200 lives, including well over 100 Americans.

Responding to public apprehension about the submarine peril and to the President's request for national defense plans, Secretary of the Navy Josephus Daniels announced, in July 1915, the formation of the Naval Consulting Board. He got the idea for such a panel from renowned inventor Thomas A. Edison who, in a newspaper interview, made an appeal for preparedness in terms of better application of American research, inventiveness, and industry to defense needs. In developing his proposal for the Board, Daniels consulted with Edison, and subsequently named him to chair it. He designated one other member—Edison's chief engineer—and then asked the 11 largest engineering societies in the country to each nominate two individuals to serve on the Board, for a total of 24 members. The panel received statutory recognition and direct funding in August 1916.[5] Subsequently, the War Department also utilized the services of the Board, and by action of February 15, 1917, the Council of National Defense—a forerunner of the National Security Council composed of a subset of Cabinet secretaries—made the panel its Board of Inventions. All of this occurred, of course, prior to American entry into the war in April 1917.[6]

The Board's work mainly consisted of evaluating suggestions, inventions, and devices offered to the federal government, devising its own solutions to mechanical problems vexing the armed forces, and otherwise offering certain expert advice.[7] Two bulletins on submarines and related technology were issued before the November 1918 armistice, as well as one concerning airplane improvement.[8] The panel also proposed the establishment of a naval research laboratory, directed by a "naval officer of rank," and staffed with "civilian experimenters, chemists, physicists, etc." Regarding one aspect of the facility's operations, the Board was unequivocal: "Secrecy should be a governing factor."[9] Funds for the project were appropriated in August 1916, but a dispute over the location of the laboratory delayed its establishment until 1923.[10] In the aftermath of the war and demobilization, the Naval Consulting Board was continued as an agency of the Navy Department.[11]

Federal interest in aeronautics—including both powered and unpowered heavier-than-air and lighter-than-air craft—during the World War I era also resulted in new organizational mechanisms where officials became acquainted with the military value of science and technology. The war actually prompted a sense of urgency for increased government attention to aviation. When hostilities ensued in Europe in 1914, France had 1,400 airplanes, Germany had 1,000, Russia had 800, and Great Britain had 400, whereas the United States counted only 23.[12] Moreover, the early unarmed observation planes were soon transformed into important reconnaissance craft; deadly fighters made their debut; and zeppelins brought the war to London—the first successful attack on British soil since 1066.

The National Advisory Committee for Aeronautics, the first federal entity created to assist with the development of an aviation service, was actually a reconstituted panel that was taken over from the Smithsonian Institution regarding its Langley Aerodynamical Laboratory.[13] According to the Committee's establishment provision in a 1913 naval appropriations act, the President was

> authorized to appoint not to exceed twelve members, to consist of two members from the War Department, from the office in charge of military aeronautics; two members from the Navy Department, from the office in charge of naval aeronautics; a representative each of the Smithsonian Institution, of the United States Weather Bureau, and of the United States Bureau of Standards; together with not more than five additional persons who shall be acquainted with the needs of aeronautical science, either civil or military, or skilled in aeronautical engineering or its allied sciences.

Serving without pay, Committee members were "to supervise and direct the scientific study of the problems of flight with a view to their practical solution, and to determine the problems which should be experimentally attacked and to discuss their solution and their application to practical questions."[14]

Although development of the Committee's own research facility at Langley Field, Virginia, did not progress very far before the armistice,[15] the panel was an important point of entree for scientists interested in the government's aviation research and development endeavors.[16] Among its other accomplishments during the war were publication of "a large number of scientific memoirs or papers dealing with the technical problems of aviation," analysis of aircraft construction and operation problems, and assistance with aircraft production and testing plans and aviator training.[17] The Committee functioned until 1958, when it was superseded and absorbed by the National Aeronautics and Space Administration.[18]

Development of a third organizational mechanism contributing to officials' acquaintance with the military value of science and technology returns attention to George Ellery Hale. A respected astronomer and an ardent promoter of scientific enterprises, Hale, as early as 1910, became an active leader among his professional colleagues seeking to rehabilitate the National Academy of Sciences to make it a true counterpart of the French Academie des Sciences and the British Royal Society.[19] Chartered by Congress in 1863, the Academy was to "investigate, examine, experiment, and report upon any subject of science or art" when requested to do so by any federal department.[20] During its first 50 years, however, the Academy, due to the rise of scientific bureaus within the executive branch, came to be ignored by the federal government. Its services were sought on only 51 occasions–"only four of them had been in the twentieth century, the last in 1908."[21] Knowledgeable of the ways of the European academies, Arthur Schuster, secretary of the Royal Society, warned his American colleagues in 1913 against being content with such exclusive, honorific status, saying "while the older institutions may offer you much that deserves to be admired and perhaps be imitated, you must not mistake the signs of gray hairs for the stamp of an enviable dignity."[22]

With the commencement of warfare on the continent, Hale began seeking improvement of the nation's defense preparedness, combining his efforts at rejuvenating the Academy with these new endeavors. By July 1915, he was advising the Academy's president that, in the event the country went to war, the Academy should offer its services to the White House–a bold and unprecedented reversal of the procedure by which the Academy came to serve the government.[23] In April 1916, Hale moved a resolution, which was unanimously accepted, that the president of the Academy "inform the President of the United States that in the event of a break in diplomatic relations with any other country the academy desires to place itself at the disposal of the Government for any service within its scope."[24] A week later, President Wilson accepted this offer and asked the Academy to proceed with developing implementing arrangements. Reaching beyond its own membership and structure, the Academy responded by creating "a National Research Council, whose purpose shall be to bring into cooperation existing governmental, educational, industrial, and other research organizations."[25] In July, a presidential letter approved the establishment of the Council, agreed to the appointment of government representatives to serve on it, and pledged department and agency cooperation with its work.[26] Subsequently, on May 11, 1918, Wilson issued E.O. 2859, which, in requesting the Academy to perpetuate the Council, specifying its duties, and promising continued government cooperation with it, permanetized the new body.[27] The Academy responded by giving the Council a fixed organization in February 1919.[28]

For his efforts, Hale was named the chairman of the Council, and served in that capacity until 1919 when he was given an emeritus status and designated its honorary chairman. His approach in leading the Council was that "the work to be undertaken should not be confined to the promotion of researches bearing directly upon military problems, but that true preparedness would best result from the encouragement of every form of investigation, whether for military and industrial application or for the advancement of knowledge without regard to its immediate practical bearing."[29] In its effort to identify and organize the scientific and technological resources of the country in the interest of national defense, the Council created a number of committees that included a broad array of representatives from the scientific and industrial communities. Of these, an active and important panel was the Research Information Committee, which was responsible for "the securing, classifying and disseminating of scientific, technical, and industrial research information, especially relating to war problems, and the interchange of such information between the allies in Europe and the United States."[30] Headquartered in Washington, the Committee maintained branch offices in London, Paris, and Rome, and worked in close cooperation with the Army and Navy intelligence offices.

The Council did not have much of a central staff until the United States formally entered the war. Operating funds were another problem: Lacking an appropriation, the Council initially relied on private contributions, the Carnegie and Rockefeller Foundations being among its earliest benefactors. Soon the White House, the Council of National Defense, and the War Department offered financial support, and the Research Council counted receipts of $270,350 at the end of 18 months of operation.[31] In February 1917, the Council of National Defense made the Naval Consulting Board its Board of Inventions and requested the Research Council to act as its Department of Research.[32]

The National Research Council, indeed, "became an agency for mobilizing the scientific talent of the country for war purposes."[33] Moreover, it became "a central scientific agency to an extent never dreamed of by the National Academy," performing "a real function as a clearinghouse of information and a focus of scientific personnel," in the view of science historian Hunter Dupree. "Most of the great research efforts of the war," he noted, "fell at least nominally within its sphere."[34]

The World War I era, however, was also a time of significant government encroachment on the First Amendment rights of all Americans.[35] Members of the scientific and technological communities were particularly affected. Censorship was an initial limitation. Shortly after the declaration of war against Germany, President Wilson, responding to a joint recommendation of the Secretaries of State, War, and Navy, issued E.O. 2594 on April 13, 1917, creating the Committee on Public Information. With a membership composed of the three Secretaries and a civilian—former newspaperman George Creel—as chairman, the Committee was responsible for both dispensing publicity about America's war effort and censorship or control of the publication of war information within the United States. In pursuit of this latter duty, the Committee issued on May 28, 1917, a "Preliminary Statement to the Press of the United States," which established voluntary censorship standards for the nation's serial publications.[36]

More imposing censorship controls, however, were in the making. As Congress approved the declaration of war on April 6, the President ordered the Secretary of the Navy to assume superintendence over all means of radio communication within the United States.[37] Three weeks later, on April 28, Wilson issued E.O. 2604, which subjected all telegraph, telephone, and submarine cable traffic to and from the country to War and Navy Department regulation.

Next came powerful new statutory weapons. The Espionage Act of June 15, 1917, provided a basis for regulating or controlling various types of expression. Section 3 of Title I, for example, made it punishable—a $10,000 fine or 20 years of imprisonment or both—to "willfully make or convey false reports or false statements with intent to interfere with the operation or success of the military or naval forces of the United States or to promote the success of its enemies."[38] Title III made it similarly punishable to "endanger the safety" of American vessels, their cargo, or persons on board.[39] Under Title VII, it was unlawful to export "any article or articles" mentioned in a presidential proclamation prohibiting such shipment, unless undertaken in accordance with prescribed regulations.[40] Title XII set restrictions on the use of the mails if the material in question was determined to be in violation of any of the provisions of the act or otherwise was found to contain "any matter advocating or urging treason, insurrection, or forcible resistance to any law of the United States."[41]

The Trading with the Enemy Act of October 6, 1917, contained similar restrictions. Section 3 of the statute, in addition to establishing broad export controls, specified:

> Whenever, during the present war, the President shall deem that the public safety demands it, he may cause to be censored under such rules

> and regulations as he may from time to time establish, communications by mail, cable, radio, or other means of transmission passing between the United States and any foreign country he may from time to time specify, or which may be carried by any vessel or other means of transportation touching at any port, place, or territory of the United States and bound to or from any foreign country.[42]

Furthermore, section 19 of the act made it "unlawful for any person, firm, corporation, or association, to print, publish, or circulate, or cause to be printed, published, or circulated in any foreign language, any news item, editorial or other printed matter, respecting the Government of the United States, or of any nation engaged in the present war, its policies, international relations, the state or conduct of the war, or any matter relating thereto," unless, prior to mailing, "a true and complete translation" was "filed with the postmaster at the place of publication."[43]

Finally, the Sedition Act of May 16, 1918 made several kinds of expression punishable, including causing or attempting to cause, or inciting or attempting to incite "insubordination, disloyalty, mutiny, or refusal of duty, in the military or naval forces of the United States"; uttering, printing, writing, or publishing "any disloyal, profane, scurrilous, or abusive language about the form of government of the United States," the Constitution, the military or naval forces, the flag, or the uniform of the Army or Navy, or using language bringing any of the same into "contempt, scorn, contumely, or disrepute"; using language promoting the cause of the enemy; or otherwise urging, inciting, or advocating "any curtailment of production . . . essential to the prosecution of the war." The statute also gave the Postmaster General authority to deny the use of the postal system to "any person or concern . . . using the mails in violation of any of the provisions of this Act."[44]

On October 12, 1917, President Wilson issued E.O. 2729A, which established the Censorship Board. Composed of representatives of the Secretary of War, the Secretary of the Navy, the Postmaster General, the War Trade Board, and the chairman of the Committee on Public Information, the Board was tasked with implementing the authority conferred on the President by section 3 of the Trading with the Enemy Act, and was "authorized to take all such measures as may be necessary or expedient to administer the powers hereby conferred." The executive order also made the Postmaster General responsible for the administration of section 19 of the Trading with the Enemy Act.[45] Together with the Attorney General, these agents of censorship proved to be indomitable.

Among members of the American scientific community, the vast majority were probably more aware of wartime censorship in their capacity as citizens than as scientists and engineers. Receipt of certain foreign professional literature might have been affected, and communication with some overseas colleagues may have been made more difficult. Furthermore, in view of Germany's leadership position in science and technology and Dupree's comment that, during the World War I era, "American science was profoundly dependent on European science in general and on Allied military research in particular," there is a serious question as to the extent to which U.S. scientific and technological knowledge were militarily valuable to the Central Powers.[46] Policywise, export controls and government security measures for the protection of information in federal possession were probably the most effective deterrents to enemy acquisition of especially useful American science or technology. Still, Creel and the Censorship Board carefully scrutinized technical periodicals, such as *Motor Age* and *Air Service Journal*, for leakage of militarily significant information.[47] An issue of *Scientific American* was held up because it contained an article giving the capabilities of a machine for straightening and checking the bore of rifle, shotgun, and machine gun barrels.[48]

However, if publications containing sensitive scientific or technological information slipped past the Censorship Board, there was the very strong possibility that its efforts might be effectively reinforced by the British. In this regard, by March 1918, for example, British censors indicated that 92 American trade and technical publications bound for European destinations via England—issues of *Motor Age, Iron Trade Review*, and *World's Work*, among others—had been barred.[49]

With the conclusion of World War I, federal censorship in the interest of national defense ceased. It would be instituted again with the entry of the United States into World War II, and a pale version would be attempted with the onset of the Cold War. Other government controls on scientific and technological communication that were statutorily inaugurated during the World War I era would persist and be refined during the ensuing years.

CENSORSHIP AND SUPPRESSION

On Sunday morning, December 7, 1941, Japanese aircraft attacked American military and naval installations at Pearl Harbor in Hawaii. The engagement lasted approximately 2 hours, and it marked the end of the United States' official stance of neutrality regarding world warfare, which had begun in the autumn of 1939. Resolution of the Pacific conflict would come 4 years later with the arrival of the atomic age. Simultaneous with the raid on Oahu, the Japanese launched assaults on the Philippines, Guam, and Midway Island. The following day, war with Japan was formally declared.[50] Three days later, the United States extended the declaration to Germany and Italy.[51]

Eleven days after the Pearl Harbor attack, the First War Powers Act was signed into law.[52] It conferred on the President authority to censor all communications from the United States with foreign countries. Domestically, the press and radio, including magazines and specialty periodicals, were controlled on a strictly voluntary and extralegal basis under a Censorship Code issued by the Office of Censorship, which was established by E.O. 8985 of December 19, 1941, pursuant to the First War Powers Act.[53] American scientific publications adhered to the new Code during the war years.

A similar arrangement, however, was in operation within the scientific community well before the declaration of war. Growing concern in 1939 among some scientists about the open publication of scientific findings of potential military value, particularly in the area of uranium fission, prompted a concerted effort by a small but influential group of scientists to seek a means to restrict the general availability of such information. Eventually, they prevailed upon the National Research Council to form a special control committee in April 1940. Through the voluntary cooperation of the editors of various scientific and technical professional journals, this panel or one of its subcommittees reviewed manuscripts and papers for possible vital defense information. Judgments were offered as to the advisability of publication. Although this review system was nominally in operation until June 1945, the rapid absorption of many scientists into government-sponsored war work of one kind or another soon almost eliminated the flow of submissions.[54]

With the surrender of Japan in August 1945, censorship as a wartime practice ceased, but interest in it or some variant continued, particularly with the onset of the Cold War. During the spring of 1948, for example, Secretary of Defense James V. Forrestal–anxious about the danger to the country posed by the premature publication of vital scientific and technical information–made a public plea for the cooperation of the press "in voluntarily refraining from publishing information detrimental to the national security." Urging the establishment of an information advisory unit within the Pentagon to answer media inquiries and offer guidance, Forrestal said he was "mainly concerned with the security of technical information, facts about new developments and weapons, about new military techniques, the knowledge of which would be of value to hostile or possibly hostile powers, and thus detrimental to the future security of this country."[55]

A media group, formed to respond to the Secretary's proposal, soon indicated it did "not believe that any type of censorship in peacetime is workable or desirable in the public interest."[56] In the face of this opposition, Forrestal abandoned the idea.[57]

In November 1954, Secretary of Commerce Sinclair Weeks announced that, at the direction of the President and on the recommendation of the National Security Council, he was creating an Office of Strategic Information within his department.[58] The mission of this new entity, according to the Secretary, was to work with various private sector groups "in voluntary efforts to prevent unclassified strategic data from being made available to those foreign nations which might use such data in a manner harmful to the defense interests of the United States."[59] The Office, however, was something of an anomaly. It had no legislative charter and its activities, in many regards, appeared to overlap with and duplicate certain more clearly and statutorily stated functions of other agencies. Because the concept of "strategic information" was not clearly defined, its regulatory application seemed to some critics to be of uneven or sometimes unfair impact. Nevertheless, the new entity was created to discover any imbalance favoring the Communist bloc in exchanges of scientific, technical, and economic information, and to alert federal agencies, as well as scientists, businesses, and the press to the dangers of indiscriminate publication of unclassified information of possible benefit to an enemy nation.

Journalists and scientists took their protest of the new Office to the House Special Subcommittee on Government Information.[60] Having serious reservations about the Office, the Subcommittee urged its abolition, a recommendation endorsed by the parent Committee on Government Operations.[61] In April 1957, the House of Representatives eliminated all funds for the Office of Strategic Information and prohibited the transfer of any money from other sources for its continuation.[62] When the Senate agreed to this action, Secretary Weeks was forced to abolish the Office.[63]

Such control arrangements do not appear to have been advanced again until 1978, when Admiral Bobby R. Inman, in a rare interview for a director of the secretive National Security Agency, told *Science* magazine that he was seeking a "dialogue" with the academic community over cryptography and computer security studies and the circumstances under which research in these areas should be protected for reasons of national security. This exclusive interview occurred in the aftermath of a series of four controversial incidents in which the National Security Agency or its employees attempted to restrict or limit the dissemination of research information. Admiral Inman said he wished to explore the possibilities of extending his agency's legal authority, without impinging on academic or other freedoms, to control privately held research information having national security implications.[64]

Responding to Admiral Inman's expressed desire for a dialogue with the academic community, the American Council on Education created a nine-member Public Cryptography Study Group in February 1980. Funded by a National Science Foundation grant, the panel examined the possible need for some restrictions on the publication of cryptographic research for reasons of protecting national security. The Study Group issued a report a year later, agreeing with the National Security Agency "that some information contained in cryptology manuscripts could be inimical to the national security of the United States" if published, rejecting the Agency's suggestion for a statutory system of prior review of articles and monographs relating to cryptology, and, with one dissent, recommending "that a voluntary system of prior review of cryptology manuscripts be instituted on an experimental basis."[65] The following guidelines were offered for the proposed voluntary arrangement.

1. NSA would notify the cryptologic community, including authors and publishers, of its desire to review manuscripts concerning cryptology prior to publication.
2. NSA, in consultation with appropriate technical societies, would define as precisely as possible those aspects of cryptology to be covered by the procedure.
3. NSA would invite authors to send manuscripts to NSA for review prior to publication.
4. NSA would assure prompt review by its staff of submitted manuscripts and prompt response to authors with an explanation, to the extent feasible, of proposed changes, deletions, or delays in publication, if any.
5. NSA would provide, in the case of unresolved disagreements, the opportunity for authors to obtain prompt review by an Advisory Committee of five persons (two appointed by the Director of NSA and three appointed by the Science Adviser to the President from a list of nominees provided by the President of the National Academy of Sciences), which would make a recommendation to the Director of NSA and to the author concerning the matters in issue. Members of the Advisory Committee shall have adequate clearance so that the committee can make informed recommendations.
6. There would be a clear understanding that submission to the process is voluntary and neither authors nor publishers will be required to comply with suggestions or restrictions urged by NSA.[66]

An April 1981 press release by the American Council on Education, formally announcing the issuance of the Study Group's report, also contained a statement by General Lincoln D. Faurer, director of the

National Security Agency, indicating acceptance of the panel's recommendations, a willingness to implement them, and the "hope that affected researchers, academicians, writers and publishers will work with us in this endeavor."[67] The Agency announced the initiation of the voluntary prepublication review program in March 1982.[68]

Dissenting from the position taken by his colleagues serving on the Study Group, George I. Davida, a professor at the University of Wisconsin at Milwaukee, submitted a minority report in which he argued against restraints. In particular, Davida contended that the restraints would be unconstitutional and implied that, before the Supreme Court rendered its decision, the restraints would promote a good many legal entanglements. Further, he maintained, such restraints would diminish the quality and direction of basic research in computer science, engineering, and mathematics, and, in addition, would not achieve the National Security Agency's objectives. Moreover, the restraints might produce international complications, given the interests that editors of foreign journals have in U.S. information. He denied that basic research results were at all likely to jeopardize the federal government's cryptosystems. Suggesting that the scope of the restrictions greatly exceeded the needs of the National Security Agency, Davida ended his dissent, saying, "I find N.S.A.'s effort to control cryptography to be unnecessary, divisive, wasteful, and chilling. The N.S.A. can perform its mission the old-fashioned way: Stay ahead of others."[69]

In June 1981, the National Science Foundation's Advisory Committee for Mathematical and Computer Sciences, which counsels the Foundation about support of cryptological research, issued a report taking specific exception to the recommendation of the Public Cryptography Study Group, calling it "unnecessary, unprecedented, and likely to cause damage to the ability and willingness of American research scientists to stay at the forefront of research in public sector uses of cryptography."[70] As an alternative to the Study Group's proposal, the Advisory Committee suggested that cryptography research papers be sent to the National Security Agency on an information basis only, not for approval or disapproval. If the Agency determined that a particular manuscript should be protected for reasons of national security, it would be incumbent upon N.S.A. to achieve that end through whatever means were legally available.[71]

By the end of 1981, Jonathan Knight, associate secretary of the American Association of University Professors, observed that the Public Cryptography Study Group's "recommendation has raised many questions among members of the academic and scientific community and of the Association's Committee A on Academic Freedom and Tenure."[72]

He indicated that Committee A had "expressed serious reservations over the ramifications of the Study Group's recommendations for academic freedom."[73]

Speaking as a panelist at the January 1982, annual meeting of the American Association for the Advancement of Science, Admiral Inman, who had recently left the directorship of the National Security Agency to become the Deputy Director of Central Intelligence, urged that the volunteer review arrangement recommended by the Public Cryptography Study Group be extended to various areas of research, including, in his view, computer hardware and software, electronic gear and techniques, lasers, crop projections, and manufacturing processes. He also suggested that such evaluation could occur "prior to the start of research and prior to publication," a most significant expansion of the Study Group's proposal.[74]

Almost three months later, testifying at a congressional hearing on the impact of national security controls on science and technology, Inman stated that increasing counterintelligence efforts by the federal government had curtailed and increasingly would continue to limit the acquisition, through illegal sales, theft, and espionage, of much American technology lately targeted by Communist bloc intelligence organizations. As a result, however, new, less protected sources of scientific and technical knowledge were being pursued. Inman said, "since the late 1970s, there has been an increased emphasis by these hostile intelligence services on the acquisition of new Western technologies emerging from universities and research centers."[75]

Consequently, Inman testified, he had sought to encourage scientists to cooperate in finding ways to improve protection for research information:

> From my past experience, I was persuaded, and I remain persuaded, that if we can energize the academic and scientific communities to think about the problem, we have a reasonable prospect that they will come up with innovative ideas which will not close off that flow, but will substantially lower the risks they, as citizens of this country, share with the rest of us.[76]

Inman, however, also appeared to be suggesting that the opportunity for voluntary controls was slipping away, saying that, as "knowledge of Soviet successes" in acquiring American scientific and technological information became better known, he worried about "what other branches of the government might ultimately propose in the way of regulation." Although, in his view, the "intelligence community will have essentially no role in writing the regulations," he said that "I think six months, a year, 18 months down the road, as the full magnitude of the Soviet's success of acquiring technology in the West comes to the front of the agenda to be considered by the Government, and the Government decides how to react, at that point I believe there will be proposals to try to regulate it." Asked to identify the formulators of such policy, Inman said: "I think it may come from a combination of Commerce, Defense, State, the entire national security apparatus."[77]

SECURITY CLASSIFICATION

A sophisticated information security classification system, affording federal records varying degrees of official secrecy, also emerged during the World War I era. Among the kinds of government information that might be protected under these arrangements were scientific and technological knowledge having national defense value.

The practice of assigning certain sensitive official records a secret status—that is, using security labels or markings—appears to have been followed within the diplomatic corps and the armed forces from the first days of the Republic, a carryover custom from colonial times and the Confederation experience. However, during the early years of the federal government, these actions were taken without explicit statutory authorization. Laws establishing the Departments of State and War, among others, contained provisions mandating the issuance of house-keeping regulations concerning the custody, use, and preservation of their papers, but no allowance was made for protecting foreign affairs or defense documents and communiques.[78] Over a half a century later, in 1857, the President was empowered "to prescribe such regulations, and make and issue such orders and instructions, not inconsistent with the Constitution or any law of the United States, in relation to the duties of all diplomatic and consular offices, the transmission of their business, . . . the safekeeping of the archives, the public property in the hands of all such officers [and] the communication of information . . . from time to time, as he may think conducive to the public interest."[79]

It appears that formal armed forces security-secrecy instructions were not issued until 1869. Initially limited to prohibiting the unauthorized photographing or sketching of forts or coastal defenses, these general orders passed through a series of metamorphoses, leading to a War Department directive of February 16, 1912, for the protection of national defense documents and papers. Certain classes of records were identified that were to be regarded as "confidential" and, therefore, kept under lock, "accessible only to the officer to whom intrusted." Materials falling into this category included submarine mine projects and land defense plans. "Trusted employees" of the War Department, as well as "the officer to whom intrusted," might have access to "maps and charts showing locations on the ground of the elements of defense, of number of guns, and of the character of the armament" and "tables giving data with reference to the number of guns, the character of the armament, and the war supply of ammunition."

Serial numbers were to be issued for all such "confidential" information, with the number marked on the document(s) and lists of the records kept at the office from which they emanated. No security labels were used. Within one year's time, officers responsible for the safekeeping of these materials were to check on their location and status. Although available to all commissioned officers at all times, "confidential" information was not to be copied except at the office of issue.

Five and one half years later, as American officers began working with their British and French allies, they had an opportunity to observe the information security arrangements of other armies. From this experience came an American Expeditionary Force general order of November 22, 1917, establishing a three-tiered security classification system. Need to know access restrictions and personnel security clearances would soon follow. War Department instructions of December 14, 1917, introduced the new A.E.F. security arrangements to home forces and imposed their use throughout U.S. military units. Moreover, improperly disclosing classified information or misusing it was made "a breach of official trust" subject to punishment under the Espionage Act.[80]

With the return to "normalcy," armed forces regulations governing the creation and safeguarding of official secrets were continued and soon assumed a pervasive character. By 1936, Army classification instructions appeared to embrace not only sensitive military matters, but also foreign policy material and what might be properly described as "political" data. The *Secret* designation covered information "of such a nature that its disclosure might endanger the national security, or cause serious injury to the interests or prestige of the Nation, an individual, or any government activity, or be of great advantage to a foreign nation." Similarly, *Confidential* could be applied to material "of such a nature that its disclosure, although not endangering the national security, might be prejudicial to the interests or prestige of the Nation, an individual, or any government activity, or be of advantage to a foreign nation." Finally, the term *Restricted* might be used in instances where information "is for official use only or of such a nature that its disclosure should be limited for reasons of administrative privacy, or should be denied the general public."

By the time of the initial years of the New Deal, armed forces security classification requirements and controls had permeated not only many civilian sectors of the War and Navy Departments, but also some other federal entities to which protected records had been transmitted. Although the creation of official secrets largely was limited to the national defense community, this authority could be exercised with broad discretion to embrace matters well beyond traditional military considerations. In addition, Army and Navy security classification regulations during World War I and into the 1930s often made general reference to criminal law to give them force, even though the cited statutory authority—the Espionage Act—seems to have been of more limited or narrow applicability than was implied by the armed forces instructions.

Security classification was effective for protecting government information, including scientific and technological knowledge, in documentary form. Engineering specifications and performance data for a new military aircraft, for example, could be readily safeguarded through these official secrecy arrangements. Shielding the actual machine, however, was another matter. When the Army brought its latest pursuit plane to the National Aviation Show in New York City during the winter of 1937, the aircraft was suspended at just above head level and then draped in yards and yards of gauze, making it only dimly visible. This precaution was taken because the airplane was a secret. The Army, nonetheless, had brought the aircraft to the show for reasons of "good publicity." Later, the gauze was removed, but visibility into the cockpit of the plane was impeded by soap applied to its windows and windshield—again, to maintain secrecy. Signs were posted requesting visitors "not to ask the guards too many questions."[81]

Security classification policy assumed a presidential character in 1940. Although the reasons for this development are not entirely clear, it was probably prompted by desires to clarify the authority of civilian personnel in the national defense community to create official secrets, to establish a broader basis for protecting military information in view of growing global hostilities, and to better manage a discretionary power of seemingly increasing importance to the entire executive branch.

Relying upon a 1938 statute concerning the security of armed forces installations and equipment and "information relative thereto," Franklin D. Roosevelt issued the first presidential security classification directive in March 1940.[82] E.O. 8381 authorized the use of control labels on "all official military or naval books, pamphlets, documents, reports, maps, charts, plans, designs, models, drawings, photographs, contracts or specifications which are now marked under the authority of the Secretary of War or Secretary of the Navy as 'secret,' 'confidential,' or 'restricted,' and all such articles or equipment which may hereafter be so marked with the approval or at the direction of the President." The order made no reference to penalties or to sanctions under the espionage laws.

For the most part, E.O. 8381 paralleled armed forces regulations for marking and handling secret records, gave civilian employees of the War and Navy Departments authority to classify information, and was confined largely to traditional national defense–rather than national security–matters.[83] However, the legislative history of the 1938 statute on which the President relied to issue his order provided no indication that Congress anticipated or expected that such a security classification arrangement would be created.

During World War II, various prerogatives were exercised to protect sensitive information. In addition to the President's order and prevailing armed forces directives, the Office of War Information, in September 1942, issued a government-wide regulation on creating and managing classified materials.[84] Other ad hoc arrangements were developed for the Manhattan Project, which is discussed shortly.

Again relying on the 1938 armed forces installations and equipment security statute, President Truman issued a new security classification directive in 1950. Superseding E.O. 8381, the new order, E.O. 10104, made only one policy change–the addition of a fourth designation, *Top Secret*–that brought United States information security categories into alignment with those of our allies.[85]

The security classification program was completely overhauled the following year by a sweeping new presidential directive. In issuing E.O. 10290, Truman indicated he was relying on "the authority vested in me by the Constitution and statutes, and as President of the United States." Four levels of classification were prescribed, but their use was no longer confined to traditional national defense matters. This policy shift was evident in two ways: First, information was now classified in the interest of "national security"; second, classification authority was extended to nonmilitary agencies, those that presumably had a role in "national security" policy. The order also resorted to citing portions of the Espionage Act to reinforce its requirements. Classified materials furnished to authorized persons "in or out of Federal service, other than the Executive Branch," were to bear a warning statement concerning their improper disclosure and giving reference to two sections of espionage law.

Criticism of the sweeping authority conveyed by E.O. 10290 from the press and science communities prompted President Eisenhower to seek a review of the directive by the Attorney General for purposes of revising or rescinding it. The subsequent recommendation was for a new order that was issued in November 1953 as E.O. 10501.[86] This executive order, later amendments to it, E.O. 11652 of 1971,[87] and E.O. 12065 of 1978[88] narrowed the bases and discretion for assigning official secrecy to agency records. Moreover, in July 1970, a Defense Science Board Task Force on Secrecy, chaired by physicist Frederick Seitz, issued a report that, among other conclusions, stated: "It is unlikely that classified information will remain secure for periods as long as five years, and it is more reasonable to assume that it will become known to others in periods as short as one year."[89] The "information" being referred to by the Task Force was scientific and technological knowledge that potentially could be independently discovered by other scientists and technologists regardless of official secrecy barriers. The report seemed to be suggesting, for the first time, that security classification was not an absolute safeguard for militarily valuable scientific or technological knowledge. The revelation raised important questions regarding the longevity of security classification protection and the utility of maintaining long-term restrictions.

In April 1982, a new order by the Reagan Administration reversed the limiting trend by expanding the categories of classifiable information, mandating that information falling within these categories be classified, making reclassification authority available, admonishing classifiers to err on the side of classification, and eliminating automatic declassification arrangements. The directive, E.O. 12356, also had implications for government restriction of scientific communication.

Because of its emphasis on vigorous classification efforts and both the increased number and breadth of its classification categories, E.O. 12356 raised the potential for more frequent and widespread application of official secrecy to scientific research produced by and for the federal government. Among the categories of information that "shall be considered for classification" under the order are "scientific, technological, or economic matters relating to the national security" and the less specific area of "vulnerabilities or capabilities of systems, installations, projects, or plans relating to the national security." In addition, for the first time, such an executive order warned "contractors, licensees, and grantees" that they shall be subject to appropriate sanctions for improperly handling classified information. As grantees, more scientists might find their government-funded studies being classified and discussions about them foreclosed to all but those who have been properly approved to received such protected information *and* have a "need to know" about the particular research in question. Access to classified information is predicated on meeting both of these requirements.[90] However, because the vast majority of the research universities will not or constitutionally cannot perform classified research, the full force of E.O. 12356 probably has been somewhat ameliorated. Consequently, from this situation, the controversy over government control of "sensitive, but unclassified" scientific information developed. Nonetheless, apart from these settings, in off-campus research facilities, security classification remains a viable restriction mechanism.

INVENTION SECRECY

Shortly after the United States formally entered World War I, Congress provided authority in one statute for the Commissioner of Patents[91] and in another–the Trading with the Enemy Act–for the President[92] to withhold certain patents, the publication of which might be "detrimental to the public safety or defense, or may assist the enemy or endanger the successful prosecution of the war," in order to keep the invention in question secret. Congress broadened the Patent Commissioner's authority in 1940 by deleting the requirement that the United States be at war in order that such protection might be exercised, and by empowering the Commissioner to withhold the grant of a patent "for such period or periods as in his opinion the national interest requires."[93] This authority lasted until the end of World War II.[94] A permanent statute on this matter was subsequently enacted in 1952,[95] and the President's parallel authority under the Trading with the Enemy Act remains available.[96]

The Invention Secrecy Act presently provides that, "[w]henever the publication or disclosure of an invention by the granting of a patent . . . might, in the opinion of the [Patent] Commissioner, be detrimental to the national security," he shall make the application available to certain specified defense agencies for review. In the event that one of these reviewing entities determines that "the publication or disclosure of the invention by the granting of a patent therefor would be detrimental to the national security . . . the Commissioner shall order that the invention be kept secret and shall withhold the grant of a patent" for not more than 1 year, subject to a possible renewal.[97] Such secrecy orders may be placed not only on inventions in which the federal government has a property interest–for example, those developed by a government employee or grantee–but also on those in which the government has no financial connection or interest, but for which, nonetheless, patent protection is sought. The act also prohibits a person from applying for a patent or similar protection in a foreign country for an invention made in the United States prior to six months after filing for a U.S. patent without a license from the Patent Commissioner. This period enables the Commissioner to determine the propriety of a secrecy order prior to the disclosure of the invention overseas.[98]

Violation of a secrecy order by unauthorized application for a foreign patent may subject a person to a maximum fine of $10,000 or imprisonment for not more than two years, or both.[99] Disclosure or unauthorized foreign patent application may also result in abandonment of the invention, constituting forfeiture of all claims against the United States based on such invention.[100] No criminal prosecutions apparently have been instituted lately for violation of the Invention Secrecy Act.[101] Much of the litigation arising from the act involves claims for compensation for damages caused by a secrecy order or the use of an invention by the government.[102] The question of the constitutionality of the statute, in terms of First Amendment constraints on the government imposing secrecy orders on privately developed information, has not been addressed by the courts, although cases adjudicating claims for compensation under the act might be viewed as impliedly upholding the statute on First Amendment grounds.[103] Finally, despite the ending of the Cold War, the number of new secrecy orders reportedly has increased steadily from 290 in 1979 to 774 in 1991, with the total number of such operative orders growing from 3,600 in 1979 to almost 5,900 in 1991.[104]

EXPORT CONTROLS

Until recent years, federal law restricted exports only in times of war or special emergencies. The Trading with the Enemy Act was the principal World War I authority in this regard.[105] In July 1940, Congress authorized the President "in the interest of national defense to prohibit or curtail the exportation of any military equipment or munitions, or component parts thereof, or machinery, tools, or material, or supplies necessary for the manufacture, servicing, or operation thereof."[106] As originally enacted, this mandate was designed to expire in only 2 years. Prosecution of the war and postwar reconversion necessitated extensions and minor adjustments of this authority.[107] The legislative history suggests that Congress regarded export controls as merely a temporary restriction on trade to avoid or mitigate the scarcity of various critical commodities and to assure their equitable distribution within the U.S. economy and to wartime allies.

The policy of exerting strict peacetime controls over United States private commercial exports to certain countries began under the duress of the Cold War with the Export Control Act of 1949.[108] This statute continued to regulate items in short supply, but also provided controls to further U.S. foreign policy goals and authorized the examination of exports to Communist bloc countries to determine if shipments contained goods having military application. In congressional testimony, American export industries expressed support for the strict new law.[109]

Administration of the statute was the responsibility of the Department of Commerce.

During the early 1950s, American scientists mailing unclassified scientific information to colleagues abroad were soon required by the Commerce Department to affix an "export license not required" label on the outside of their envelopes to indicate they were aware of export regulations, but had not violated general export license requirements. This situation resulted from a position taken by the Department that the Export Control Act conveyed authority to regulate any export, commercial or noncommercial, of commodities or technical data, with "technical data" being interpreted to include even basic scientific research information.[110]

At the time, two high-level studies had recently counseled against such restrictions. In its 1947 report, the President's Scientific Research Board, chaired by presidential aide John R. Steelman, concluded that federal security regulations "should be applied only when strictly necessary and then limited to specific instruments, machines or processes. They should not attempt to cover basic principles or fundamental knowledge."[111]

A 1950 report on the role of the Department of State in the field of national and international science, prepared for the Department by Lloyd V. Berkner under the sponsorship of the National Academy of Sciences, found "ample evidence that unnecessary restrictions exist on the flow of unclassified scientific and technological information" and, among other relevant conclusions, stated: "These restrictions are dangerous to the progress of United States science, affect adversely the conduct of our foreign relations in science, and are therefore damaging to our national security." The report also commented that, regarding "unclassified technical information, no definite and positive controls are possible under our system of government. It would be inimicable to our way of life to attempt to control the use of information that is freely available to anyone." Finally, the report warned: "The principal danger of unnecessary restrictions lies in the creation of a furtive atmosphere in which the flow of information necessary to progressive science is brought to a halt."[112]

Scientists and others brought their objections to the export regulatory efforts of the Commerce Department and related matters to the attention of the Special Subcommittee on Government Information of the House Committee on Government Operations.[113] The panel championed a recommendation, formally endorsed by its parent committee, that the Export Control Act be amended to make it clear that "technical data" did not include noncommercial, unclassified scientific information.[114] In response, the Commerce Department revised its regulations to permit overseas mailings of unclassified scientific and technical information if the material was sent first class by a person without "commercial" connections. A short time later, another revision allowed scientists to use any class of mail for their letters overseas. Still, the controversy did not subside until a more relaxed policy on exports to Communist bloc countries emerged in the 1960s and the Export Administration Act of 1969, reflecting this new position, was adopted.[115]

By 1969, Cold War attitudes toward the Soviet Union and its Warsaw Pact allies had undergone considerable transformation: Detente was the operative principle and many in Congress were supportive of a relaxation of East-West trade restrictions. The Export Administration Act of 1969 represented a watershed in United States strategic trade history with its rejection of the prior doctrine of economic warfare with the Communist bloc. Indeed, the statute's title seemed to reflect a less hostile export policy perspective. Export controls for traditional policy purposes—national security, foreign relations, and domestic short supply—were continued, but were to be lifted from goods and technologies that were freely available from foreign sources and that were only marginally of military value. In general, the law and its administration appear to have had little limiting effect on traditional scientific communication.[116]

The statute underwent some further development during the next few years,[117] then the export program was substantially overhauled after a decade by the Export Administration Act of 1979.[118] The new law sought, consistent with protecting national security and preserving some flexibility for the President to exercise foreign policy controls, to encourage trade to the greatest extent possible. However, at the time of the act's legislative formulation, détente was breaking down. Heightened concern about the Communist bloc gaining military or tactical advantages through the unauthorized acquisition of useful technology from the United States or its allies prompted support for new provisions in the act addressing this prospect. Moreover, due to subsequent Soviet actions against Afghanistan and Poland, the trade liberalization envisioned by the statute never occurred.

Among the substantial policy changes effected by the 1979 export law were the establishment of separate and distinct procedures and criteria for imposing national security and foreign policy controls. Also, precise deadlines were set for the timely processing of export license applications.

In addition to authorizing the Secretary of Commerce to establish and maintain a Commodity Control List of goods or technologies subject to export regulation, the statute mandated the development of a Militarily Critical Technologies List by the Secretary of Defense. This latter list had been recommended by a 1976 Defense Science Board task force chaired by J. Fred Bucy, Jr., who was then executive vice president of Texas Instruments.[119] To be completed and published "in an appropriate form" in the *Federal Register* by October 1, 1980, this list was to be integrated into the Commodity Control List maintained by the Secretary of Commerce. Although the Militarily Critical Technologies List was developed, only its table of contents was initially published.[120] The actual list remained security classified. An unclassified version was produced in October 1986. Moreover, integration of the contents of the Militarily Critical Technologies List into the Commodity Control List did not immediately occur, and was eventually abandoned: The two lists are separately maintained.[121]

The new statute also formally authorized United States participation in the Coordinating Committee on Multilateral Export Controls.[122] Known as COCOM, this informal multilateral export control body was secretly formulated in 1949, but met infrequently and had few resources until 1984. With a staff of only 14 and no word-processing capability in 1983, its total annual budget was less than $500,000. The following year, the Department of Defense (DOD) began upgrading the organization with a $2 million contribution for efficient information technology.[123] Today, it has a membership composed of western European nations belonging to the North Atlantic Treaty Organization, except Iceland, plus Japan and recently Australia. U.S. domination of the organization and its export policy was viewed by some member governments as an indication of American distrust of allies and often prompted compensatory assistance for national industries.[124] Still largely funded and staffed by the United States, COCOM's offices are located in Paris adjacent to the American embassy.

Shortly after the Export Administration Act of 1979 was signed into law, Pentagon officials appeared before a panel of congressional overseers to discuss their new organization for and performance in expediting export license applications.[125] At that proceeding, indications were that denying technology transfer to potential adversaries was being pursued within the context of the licensing process, suggesting that DOD was focusing its export regulatory efforts on overseas commercial trade transactions, not traditional scientific communication.

The need to reauthorize the act resulted in legislative amendments in 1985 that, among other things, eased some licensing requirements and obligated the United States to negotiate with its COCOM allies to achieve greater coordination and compliance with multilateral export controls, fewer exceptions to the mutually agreed upon list of controlled items, and strengthened and uniform enforcement.[126] Licensing adjustments included elimination of requirements for exports of certain relatively low-technology goods and technology going to destinations within the borders of COCOM member nations, and creation of a new "comprehensive operations license" to obviate the need for individual validated licenses for day-to-day technical data transactions within multinational enterprises and between other related firms certified in this regard by the Commerce Department. A procedure was also established for review and decontrol of goods found to be widely available overseas and that cannot be brought under control.[127] Efforts to relieve traditional scientific communication from the national security provisions of the Export Administration Act, however, were not very successful.

The congressional focus when reauthorizing the statute, as reflected in committee hearing transcripts, appeared to be once again on trade, foreign market, and related economic considerations. Consequently, the professional communication difficulties experienced by scientists as a result of the Export Administration Act were largely unclear to and unappreciated by Congress.

Furthermore, for whatever reasons, no scientist or representative of a scientific professional organization appeared before the congressional subcommittees and committees reauthorizing the export law to appeal for some form of relief for traditional scientific communication from national security export controls.[128] Such witnesses had expressed their distaste for these and other national security restrictions before other congressional panels, ones lacking jurisdiction to provide legislative assistance.[129]

During a House Committee on Foreign Affairs mark-up of the legislation revising and reauthorizing the Export Administration Act, Representative Lee Hamilton offered an amendment inserting the following declaration into the bill:

> It is the policy of the United States to sustain vigorous scientific enterprise. To do so requires protecting the ability of scientists and other scholars freely to communicate their research findings by means of publication, teaching, conferences, and other forms of scholarly exchange.[130]

The Committee's acceptance of the provision was subsequently explained in its report on the legislation, saying:

> The committee is deeply concerned that an overly broad interpretation of the Export Administration Act may seriously limit, on grounds of national security, the legitimate scientific communication process on which scientific productivity in the United States depends. Clearly, the strength of U.S. technology which underlies national security will not be maintained

> or improved if scientific and technological progress and innovation are inhibited as a result of overreaching security limitations on dissemination of scientific information under the Export Administration Act.[131]

Although this policy declaration remained in the reauthorization legislation adopted by the House of Representatives, it was changed during conference committee reconciliation of the measures adopted in each congressional chamber revising the export law. As finally enacted, the provision reads:

> It is the policy of the United States to sustain vigorous scientific enterprise. To do so involves sustaining the ability of scientists and other scholars freely to communicate research findings, in accordance with applicable provisions of law, by means of publication, teaching, conferences and other forms of scholarly exchange.[132]

The conference committee report did not explain the significance of the modification.[133] However, the addition of the qualifying phrase—"in accordance with applicable provisions of law"—would appear to diminish or defeat the original purpose of the declaration if it can be maintained that, among the "applicable provisions of law" referenced in the new section are those of the Export Administration Act itself.

The 1985 reauthorization act continued the Export Administration Act, as amended, to the end of September 1989,[134] and this expiration date was extended, with further modification of the export law, in the Omnibus Trade and Competitiveness Act of 1988 by one year.[135] When congressional hearings were held on reauthorization legislation during the early months of 1990, witnesses from the scientific community again failed to materialize.[136] Subsequently, however, the resulting reauthorization legislation was vetoed by President Bush;[137] earlier, the regulations implementing the Export Administration Act had been continued by the President through the invocation of national emergency powers.[138]

Ultimately, efforts by scientists and their representatives to obtain legislative relief during the reauthorization of the export law may well have been overtaken and frustrated by a larger contest—a continuing constitutional rivalry between Congress and the executive over matters—foreign policy, war powers, and national security—concerning the external relations of the United States. As noted earlier, citizens, such as scientists, seeking some change in such areas can get caught up in this interbranch rivalry with the result of being manipulated within or excluded from the policymaking struggle.[139]

Congress did amend and significantly modify the Export Administration Act when adopting the Omnibus Trade and Competitiveness Act of

1988. This comprehensive and complex trade law was developed at a time of strong concern about the nation's growing trade deficit, widespread interest in improving America's competitive position in international markets, and much improved relations with the Soviet Union as a result of new leadership in the Kremlin. Exports were promoted by a number of general enhancements and by amendments to the export law eliminating licenses for shipments to COCOM countries of goods that may currently be exported to the People's Republic of China without multilateral approval, for exports to COCOM countries that are judged to have effective export control systems, and for shipments to non-Soviet bloc countries of low-technology goods that could be exported to Communist bloc nations without multilateral approval. Re-export licenses for items sold to COCOM countries also were eliminated. Furthermore, struck from the list of controlled goods was technology that could be exported to Soviet bloc countries without multilateral approval, except for items on which COCOM members agreed to maintain notification requirements. The extent to which these changes affect dual use scientific and technological knowledge, however, remains to be seen.

Another export control law, the Arms Export Control Act, has been somewhat less troublesome for traditional scientific communication. It provides for the regulation of the export of "defense articles" and "defense services" in furtherance of "world peace and the security and foreign policy of the United States."[140] Enacted in 1976, the statute, which does not require periodic reauthorization, superseded and consolidated most of the arms control provisions of the Mutual Security Act of 1954,[141] the Foreign Assistance Act of 1961,[142] and the Foreign Military Sales Act of 1968.[143] As a result, all major arms sales requests of foreign nations are channeled through the federal government, establishing a control over commercial arms and weaponry sales activity by private industries. Furthermore, the statute broadened congressional authority to review and to veto proposed arms sales agreements.

Administered by the Department of State, the Arms Export Control Act is implemented through the International Traffic in Arms Regulations.[144] They define *export* broadly to include "(s)ending or taking technical data outside of the United States in any manner except by mere travel outside of the United States by a person whose personal knowledge includes technical data" and "(d)isclosing or transferring technical data to a foreign person."[145] The term *technical data* is defined to embrace "(i)nformation which is directly related to the design, engineering, development, production, processing, manufacture, use, operation, overhaul, repair, maintenance, modification, or reconstruction of defense articles." However, it "does not include information concerning general scientific, mathematical or engineering principles."[146] Similarly, the statute's licensing requirements do not extend to "public domain" information, or that "which is published and which is generally accessible or available to the public (t)hrough sales at newsstands and bookstores . . . subscriptions which are available without restriction . . . second class mailing privileges, or . . . libraries open to the public."[147]

In brief, the statute regulates the export of goods, services, or knowledge having direct and singular military use. Far more troublesome for traditional scientific communication have been the dual use controls of the Export Administration Act, which seemingly apply to a greater variety of scientific and technological information. "Articles, services, and related technical data" subject to Arms Export Control Act licensing are generally enumerated in the United States Munitions List.[148] The civil enforcement powers and penalties under the Arms Export Control Act are the same as those under the Export Administration Act.[149] Willful violation of the statute or regulations also may subject a person to criminal penalties of a maximum $100,000 fine or imprisonment for two years, or both.[150]

ATOMIC ENERGY

Apprised in October 1939 of atomic power capabilities and the likelihood of military applications, President Roosevelt directed his science adviser, Vannevar Bush, to begin planning American development and production of an atomic bomb. Bush subsequently recommended that the Army assume design and construction responsibility for the undertaking. The President agreed and a special unit for this purpose soon was created within the Army Corps of Engineers—the Manhattan District.[151] Unlike the other Engineer Divisions and Districts, the Manhattan Project had two unique characteristics: Its operations were not confined to a single geographic locale, and it was clothed in official secrecy. General Leslie R. Groves, the designer and builder of the Pentagon, was the principal manager and supervisor of the Manhattan District almost from its inception in August 1942, until its operations were assumed by the new Atomic Energy Commission in 1946.[152]

Security arrangements for the Manhattan Project initially were handled by the Army counterintelligence command and the Federal Bureau of Investigation, but this responsibility shortly was assumed by the District's "own complete security staff" which, by Grove's account, consisted of 485 individuals by the end of the war. Liaison and cooperation with the FBI apparently continued, lasting throughout the entire course of the Manhattan program.[153]

Various procedures were used to maintain the security of the project. Foremost among these, in the view of General Groves, was compartmentalization of knowledge: "Each man should know everything he needed to know to do his job and nothing else."[154] In addition, code names and words were used to mask certain sensitive and secret matters; both official and personal correspondence was monitored for indiscretions; "loose-talk" cases were investigated and acted on; project personnel were scrutinized and required to meet established loyalty and integrity criteria; a vigorous security education program was maintained; public relations efforts were made to obtain the cooperation of newspapers and the populace of communities adjacent to project plant sites; and a censorship review program was established to limit media reports relating to the existence of the Manhattan Project, its activities or those of its component units, personnel, or contractors, or to atomic theory, equipment, and materials. The censorship review program was conducted in cooperation with the Office of Censorship and, by the spring of 1944, covered 370 newspapers and 70 magazines.[155]

Personnel having access to any official secrets of the project were subject to a background investigation to establish their loyalty, integrity, and discretion. Approved individuals were informed of the penalties for disclosing classified information improperly and "then required to read and sign either the Espionage Act or a special secrecy agreement."[156] By August 1945, some 400,000 Manhattan District employees reportedly had been investigated and approximately 600 companies involved in the project had been cleared. Business firms received security approvals if Army intelligence files, a Dun and Bradstreet credit report, and agency checks of key personnel yielded no adverse findings.[157]

In the aftermath of World War II, federal control of atomic energy research and development was given a statutory mandate that placed the program in civilian hands, but continued the stringent security and secrecy arrangements of the Manhattan Project. The rigid controls of the Atomic Energy Act of 1946[158] were relaxed somewhat in 1954 when private industry was given the opportunity to explore the applications of nuclear power.[159] Nonetheless, the law still gives the federal government exclusive authority over so-called *Restricted Data* "concerning (1) design, manufacture, or utilization of atomic weapons, (2) the production of special nuclear material, or (3) the use of special nuclear material in the production of energy."[160] Information in this category is "born classified," that is, it is subject to official secrecy protection from the moment of its creation, regardless of where located, and can be removed from its privileged status only by an affirmative determination of the government.[161] As a consequence of this situation, the federal government has moved against private parties, including both scientists and nonscientists, who allegedly have produced Restricted Data, by imposing secrecy restrictions. Although there is some question as to whether or not Congress intended the Restricted Data protections of the Atomic Energy Act to be so applied to private individuals who independently developed information asserted to be of this type, executive branch officials have maintained this position when imposing such restrictions. The issue arose several years ago when the government restrained the *Progressive* magazine from publishing an article on the way a hydrogen bomb works.[162] The Atomic Energy Act also contains provisions regulating exports, protecting nuclear power plant safety information, and safeguarding power plant facilities and nuclear materials.

CONTRACT CONTROLS

Since the end of World War II, the federal government has been a major source of funding for scientific research, particularly for university-based scientists, and in the past decade, the Department of Defense has controlled the predominant portion of these funds. Moreover, although most universities constitutionally cannot or otherwise are reluctant to undertake classified research, many of them are willing to allow federal research contracts requiring signatories' agreement to submit results deriving from unclassified government-sponsored research for prepublication review. Although this practice has First Amendment implications, no court decisions involving prepublication review clauses in contracts and grants to universities have been reported.[163]

Also of pertinence are Pentagon directives proscribing professional and academic meeting presentation of unclassified research papers containing export-controlled information unless dissemination and access controls are "equivalent to those used for distributing the data directly by DoD."[164] Such regulation was reinforced in 1983 by statutory authority empowering the Secretary of Defense to "withhold from public disclosure any technical data with military or space application in the possession of, or under the control of, the Department of Defense, if such data may not be exported lawfully outside of the United States without an approval, authorization, or license."[165] This provision has lent support to the creation of an information subsystem for unclassified, export-controlled technical data through the use of special distribution agreements, distribution statements, and withholding regulations.[166] In furtherance of such arrangements, the Patent and Trade Mark Office in September 1986 revised the scope of secrecy order use by defense agencies by creating three new types.[167] One of these may be applied by the Patent Commissioner when the Pentagon determines that unclassified technical data in a patent application should be subject to export controls. However, under the terms of this order, it is permissible to disclose protected information for business purposes to American citizens within the United States. Furthermore, foreign patent applications containing protected information may be made in Australia, Canada, and most of Europe, but not in Japan or other newly industrializing or developing countries having no comparable patent secrecy authority.

Another contractual innovation emerged in the spring of 1983. A national security decision directive, NSDD 84, entitled "Safeguarding National Security Information," was publicly released by the Reagan Administration in mid-March. Although signed by the President, such directives, unlike executive orders, are not required to be published in the *Federal Register*. They are a specialized presidential instruction series, usually assigned a security classified status, and maintained in the files of the National Security Council.[168] NSDD 84 provided that:

- All present and future employees of the executive branch with authorized access to classified information be required to sign a nondisclosure agreement as a condition of access to such information.
- All present and future employees of the executive branch with authorized access to sensitive compartmented information or SCI, a kind of intelligence information, be required to sign a nondisclosure agreement as a condition of access to SCI and other classified information, with a provision mandating prepublication review of public texts and manuscripts for publication to assure deletion of SCI and other classified information.
- Executive branch employees be required to submit to polygraph examinations, when appropriate, in the course of investigations of unauthorized disclosure of classified information, including SCI.
- The departments and agencies develop and adopt appropriate policies and procedures to govern contacts between media representatives and executive branch employees, so as to reduce the opportunity for negligent or deliberate disclosures of classified information.[169]

Secrecy agreements of the type mandated by NSDD 84 had been in use for some time within the federal intelligence community and a few other agencies possessing particularly sensitive information.[170] In 1980, such agreements were found by the Supreme Court to be a proper enforcement device to prevent unauthorized disclosure of classified information.[171] These agreements specify a fiduciary relationship between federal employees and the government; if they are violated by an improper disclosure of information, punishment may be pursued as a breach of contract without regard to the motive for the infraction. NSDD 84 extended the use of secrecy agreements to all executive branch personnel, including contractors and grantees, having access to classified information. It also required all agency employees, contractors, and grantees outside of the intelligence community who had access to sensitive compartmented information (SCI), to submit their public writings, speech texts, and manuscripts for publication to prior review by government officials. This latter secrecy agreement would be binding upon signatories for the rest of their lives, regardless of whether or not they continued to have access to SCI.

The new directive prompted not only a number of protests against its breadth and accompanying demands, but also serious questioning of the related use—as well as the reliability and validity—of polygraph testing.[172] Various congressional panels held hearings on the need for NSDD 84 and urged delaying its implementation to allow further study of the matter.[173] Administration representatives appeared before all of these bodies to defend the ordered policy. However, because there seemed to be no willingness to postpone the effectuating of the directive, two actions were taken by Congress. First, provisions were appended to the Department of Defense authorization in both the Senate[174] and the House[175] to prevent increased use of polygraph tests by the Pentagon (which has the largest number of personnel having access to classified information), as anticipated by NSDD 84, prior to April 15, 1984.[176] This temporary moratorium remained in the authorization legislation signed by the President in late September 1983.[177]

Next, a provision was appended in the Senate to the Department of State authorization to prevent the postemployment application of the prepublication review requirement of NSDD 84 prior to mid-April 1984.[178] This prohibition became law when the President approved the legislation in late November 1983.[179]

Both of these temporary limitations on the implementation of portions of NSDD 84 provided a clear indication of congressional displeasure. The following year, legislation was introduced to make these proscriptions permanent.[180] As this bill began receiving committee consideration in the House,[181] the press reported that the Reagan Administration was willing to concede on the points of greatest objection to Congress.[182] Shortly, the President's national security adviser, Robert McFarlane, informed managers of the legislation that, on President Reagan's instructions, plans to implement the polygraph testing and postemployment prepublication review provisions of NSDD 84 were being suspended "for the duration of this Congress."[183] Consequently, no further action was taken on the pending bill.

The administration discontinued using the secrecy agreement form initially prepared pursuant to NSDD 84, requiring postemployment prepublication review of writing by signatories having access to SCI. However, as the General Accounting Office discovered, an older form utilized by the intelligence community, also containing the postemployment prepublication review requirement, was substituted.[184] To rectify this situation, a new form was then prepared, but raised issues concerning its constitutionality, legality, and wording. In response, Congress, in a joint resolution continuing appropriations for fiscal year 1988, set a qualified prohibition on utilizing appropriated funds to implement or enforce secrecy agreements of this type.[185] In May 1988, this restriction came under constitutional challenge,[186] provoking additional congressional ire.[187] The Supreme Court, however, remanded the case, on various technical grounds, to the District Court from which it arose for further consideration.[188] The dispute was eventually vacated when, in 1990, Congress and the executive agreed on "an affirmative statement" that not only was to be included in all nondisclosure forms and policy statements, but also was statutorily mandated.[189] "The purpose of the required statement," according to a conference committee report, "is to assure that the restrictions in nondisclosure policies, forms and agreements do not supersede, conflict with, or otherwise alter the employee obligations, rights or liabilities created by Executive Order 12356, on classifying and safeguarding national security information, and several statutes" specifically cited in the statement.[190] The new requirement was implemented in late January 1991.[191] NSDD 84 remains operative policy.[192]

OFFSETS

Scientists considering accepting research funds through federal contracts or grants have an opportunity to inquire about the possibilities of national security controls—prepublication review, security classification, or a nondisclosure agreement—attending such largess. The fact of such restrictions or the likelihood of their later imposition during the performance of research can be considered before research funds are accepted. However, what about national security controls—export restrictions or patent secrecy orders—imposed without warning on scientific knowledge resulting from research performed without federal support? Indeed, what assurances do American scientists have that their professional expression enjoys First Amendment protection in the face of national security controls?

As noted earlier, when discussing scientific freedom, Harold P. Green concluded that there is no "precedent or legal authority that clearly supports the proposition that there is a constitutionally protected right to pursue knowledge or to engage in scientific inquiry." In his view, "such rights are implicit in the First Amendment freedom of speech and press."[193] Another legal analyst, Steven Goldberg, concurred, saying First Amendment protection "is no more absolute for scientific speech than for any other variety."[194] Also, added attorney James Ferguson, "the Supreme Court has strongly hinted on several occasions that scientific speech is a protected form of expression."[195] However, shifting slightly from Goldberg's position, he has sought "to show not only that scientific communication is a protected form of speech, but also that it does not occupy a 'subordinate position' in the emerging hierarchy of first amendment values."[196]

Ferguson has developed "a useful analytical framework" from a 1976 Supreme Court decision concerning so-called *commercial speech*.[197] Ruling that commercial speech is not "subject to complete suppression by the State," the Court determined that such information has First Amendment "value" by identifying ways in which it promotes three major interests that are central to the free speech guarantee.[198] These Ferguson characterizes as "(1) an individual interest in self-expression, (2) a general social interest in the free flow of information, and (3) a more specific social interest in enlightened public decisionmaking."[199] He subsequently concludes "that scientific expression is so crucial to the individual concerns of scientists, so vital to the material and cultural concerns of the public-at-large, and so important to the policy concerns of a democratic state that it warrants more than the 'limited measure of protection' that the Court has extended to commercial speech."[200] As he has developed his thinking on this subject, Ferguson seems to foresee the justices of the High Court taking a cautious and balanced approach in resolving disputes between scientific expression and national security controls. Ferguson wrote:

> Indeed, if the Court applies its standard analysis to this issue, it will not only give due weight to the value of scientific freedom but will also examine critically the nature and magnitude of the threatened harm to national security. In addition, it will address a variety of other considerations, such as the technical skills of the receiving nation and the reasonableness of the regulatory technique. By incorporating each of these factors into a method of adjudication that formally allocates the burden of proof, the Court's approach provides a well-defined analytical framework

for accommodating the claims of scientific freedom with the legitimate interests of national security.[201]

Judicial practice and certain decisions of the Supreme Court concerning national security, however, do not encourage or substantiate Ferguson's relatively optimistic outlook. At issue is judicial deference to executive branch judgments on national security matters, as the adjudication of Freedom of Information Act (FOI Act) cases illustrates.

Although the FOI Act, as enacted in 1966, required *de novo* determination of agency actions by the federal courts, some ambiguity existed regarding the extent to which the courts might engage in *in camera* inspection of withheld records.[202] In 1973, the Supreme Court held that the content of documents withheld under the security classification exemption of the statute were not reviewable under the *de novo* requirement.[203] In the Court's majority view, the limit of judicial inquiry is the determination of whether or not the information was, in fact, marked properly with a security classification label under specific requirements of an executive order, and that this determination was satisfied by an affidavit from the agency controlling the information. *In camera* inspection of the records by a trial court to determine if the contested information actually fell within the criteria of the executive order was specifically rejected. Although Congress clarified the situation in 1974 with amendments to the FOI Act empowering, but not requiring, *in camera* inspection by the courts of any withheld records,[204] the courts generally have exercised this option very infrequently in cases involving withholding for reasons of national security and prefer, instead, to defer to government affidavits.[205] Indeed, in 1977, the U.S. Court of Appeals for the District of Columbia Circuit suggested that such review should be performed only when government affidavits are deficient.[206] These cases demonstrate that federal courts, when given the option of conducting the kind of cautious and balanced assessment envisioned by Ferguson, prefer, in practice, to defer to executive branch judgments about protecting national security.

Some other case law developments also do not appear to support the cautious and balanced approach foreseen for the courts by Ferguson. A useful beginning point is the Supreme Court's 1971 ruling on the government's attempt to enjoin publication of the so-called Pentagon Papers by the *New York Times* and *Washington Post*, ostensibly for national security reasons.[207] In a *per curiam* opinion, the Court found that the government had not met its heavy burden of justifying its case. Some of the justices also were disturbed that, in seeking restraint, the government, in Justice Black's words, "does not even attempt to rely on any act of Congress."[208]

However, Justice Stewart, with whom Justice White joined in his concurring opinion, appeared to be favorable to sustaining prior restraint in situations where the disclosed information "will surely result in direct, immediate, and irreparable damage to our Nation or its people."[209] Indeed, both Stewart and White conceded that some harm would result from press publication of the Pentagon Papers, but both felt that the government had not made a strong enough argument for prior restraint.[210]

Nonetheless, there is a strong likelihood that Justice Stewart's prior restraint test will be met if the courts carry over from FOI Act cases involving national security protection of information the same excessive deference they accord to assertions of fact by executive branch officials. Furthermore, it appears that the government's case for prior restraint is strengthened by invoking at least general statutory authority in support of such enjoinment.[211] In this regard, the export laws and the Invention Secrecy Act arguably mandate a national security prior restraint. And, although the litigation was later mooted, such a restraint was attained under the Atomic Energy Act in the *Progressive* case.[212] Furthermore, the question of the constitutionality of the Invention Secrecy Act, in terms of First Amendment constraints on the government imposing secrecy orders on privately developed scientific or technological information, has gone unaddressed by the courts, but litigation adjudicating claims for compensation under the statute might be seen as impliedly upholding the act on First Amendment grounds. In the case of export authority, even when the opportunity has presented itself, as it did in 1978 in the *Edler* case, the court avoided possible constitutional invalidation by adopting a narrow interpretation of the scope of the technical data controlled.[213]

The executive branch may also assert that, where national security is at stake, it has some extraordinary (or "extrastatutory") authority to impose prior restraint, and the courts, indeed, have shown some agreement with this point of view. Several years ago, for example, in the *Agee* case, the Supreme Court sustained the revocation of the passport of a former employee of the Central Intelligence Agency on the ground that his overseas activities—identifying CIA agents—caused "serious damage to the national security [and] foreign policy of the United States." Although the former employee argued that Congress had not authorized the Secretary of State to revoke passports on this basis, the Court rejected this contention. In doing so, the Court not only broke with important aspects of its own past rulings in this area,[214] but also signaled its agreement that the government's position took precedence over Agee's First Amendment freedom of speech, saying "[i]t is 'obvious and unarguable' that no governmental interest is more compelling than the security of the Nation."[215] This view does not appear to reflect a cautious and balanced assessment of First Amendment rights and national security interests.

A year before, in another case involving a former CIA employee, the Supreme Court dismissed the defendant's First Amendment defense in a footnote, saying "[t]he government has a compelling interest in protecting both the secrecy of information important to our national security and the appearance of confidentiality so essential to the effective operation of our foreign intelligence service."[216] Accordingly, the Court upheld the enforceability of the former CIA employee's agreement to submit his writings for prepublication review, even though no classified information was at issue. Thus, "the appearance of confidentiality," in the Court's view, outweighed the First Amendment interests at stake. Furthermore, from this case, the Court may be willing to approve restrictions on the national security expression of any citizen, regardless of whether or not the individual is employed by the federal government. The decision does not reflect the cautious and balanced approach envisioned by Ferguson, but does support the likelihood of Court agreement that statutorily mandated national security controls take precedence over First Amendment rights of expression where maintenance or preservation of national security is at issue.

NOTES AND REFERENCES

[1]Robert M. Yerkes, ed. *The New World of Science: Its Development During the War*. New York: The Century Company, 1920, p. xiii.

[2]A. Hunter Dupree. *Science in the Federal Government: A History of Policies and Activities*. Baltimore: The Johns Hopkins University Press, 1986, pp. 308–309; also see Daniel J. Kevles. "George Ellery Hale, the First World War, and the Advancement of Science in America." *Isis*, v. 59, Winter, 1968, pp. 427–437.

[3]See, for example, Robert V. Bruce. *Lincoln and the Tools of War*. Indianapolis: The Bobbs-Merrill Company, 1956; Nathan Reingold. "Science in the Civil War: The Permanent Commission of the Navy Department." *Isis*, v. 49, September, 1958, pp. 307–318; Merritt Roe Smith, ed. *Military Enterprise and Technological Change: Perspectives on the American Experience*. Cambridge. MA: The MIT Press, 1985.

[4]U.S. Congress. House. Committee on Science and Technology. *A History of Science Policy in the United States, 1940–1985* by Jeffrey K. Stine. Committee print, 99th Congress, 2d Session. Washington, DC: U.S. Govt. Print. Off., 1986, p. 8.

[5]39 Stat. 556, 558.

[6]40 Stat. 1.

[7]The Board reportedly examined 110,000 suggestions, but only 110 inventions merited detailed scrutiny by its subcommittees of experts, and only one of these went into actual production. Dupree, *Science in the Federal Government*, p. 308.

[8]See Naval Consulting Board of the United States. *The Submarine and Kindred Problems*. Bulletin No. 1. New York: Naval Consulting Board, 1917; Naval Consulting Board of the

United States. *The Enemy Submarine*. Bulletin No. 2. New York: Naval Consulting Board, 1918; Naval Consulting Board of the United States. *Problems of Aeroplane Improvement*. Bulletin No. 3. New York: Naval Consulting Board, 1918.

[9]Lloyd N. Scott. *Naval Consulting Board of the United States*. Washington, DC: U.S. Govt. Print. Off., 1920, pp. 109–113.

[10] *Ibid*., pp. 112–113; Dupree, *Science in the Federal Government*, p. 307.

[11]Details on the organization and operation of the Board during the World War I era may be found in Robert D. Cuff. *The War Industries Board: Business-Government Relations During World War I*. Baltimore: The Johns Hopkins University Press, 1973, pp. 13–26; Lloyd N. Scott. *Naval Consulting Board of the United States*. Washington, DC: U.S. Govt. Print. Off., 1920; William Franklin Willoughby. *Government Organization in War Time and After*. New York: D. Appleton and Company, 1919, pp. 30–32.

[12]Dupree, *Science in the Federal Government*, p. 286.

[13]See *Ibid*., pp. 283–287.

[14]38 Stat. 928, 930.

[15]Dupree, *Science in the Federal Government*, p. 318.

[16]*Ibid*., p. 297.

[17]Willoughby, *Government Organization in War Time and After*, p. 330.

[18]See Alex Roland. *Model Research: The National Advisory Committee for Aeronautics, 1915–1958*. Washington, DC: National Aeronautics and Space Administration, 1985.

[19]Dupree, *Science in the Federal Government*, p. 308.

[20]12 Stat. 806.

[21]Kevles, "George Ellery Hale, the First World War, and the Advancement of Science in America," p. 430.

[22]National Academy of Sciences. *Report of the National Academy of Sciences for the Year 1913*. Washington, DC: U.S. Govt. Print. Off., 1914, p. 86.

[23]Dupree, *Science in the Federal Government*, pp. 308–309.

[24]*Ibid*., p. 309; National Academy of Sciences. *Report of the National Academy of Sciences for the Year 1916*. Washington, DC: U.S. Govt. Print. Off., 1917, p. 12.

[25]Dupree, *Science in the Federal Government*, p. 309.

[26]*Ibid*.; Willoughby, *Government Organization in War Time and After*, p. 23.

[27]The full text of the executive order appears in Willoughby, *Government Organization in War Time and After*, pp. 23–24, and in National Academy of Sciences. *Report of the National Academy of Sciences for the Year 1918*. Washington, DC: U.S. Govt. Print. Off., 1919, pp. 40–41. On the formulation of the executive order, see Kevles, "George Ellery Hale, the First World War, and the Advancement of Science in America," pp. 432–435.

[28]National Academy of Sciences. *Report of the National Academy of Sciences for the Year 1919*. Washington, DC: U.S. Govt. Print. Off., 1920, pp. 22–26.

[29]National Academy of Sciences. *Report of the National Academy of Sciences for the Year 1916*, p. 32.

[30]Willoughby, *Government Organization in War Time and After*, p. 25.

[31]National Academy of Sciences. *Report of the National Academy of Sciences for the Year 1917*. Washington, DC: U.S. Govt. Print. Off., 1918, p. 70.

[32]Dupree, *Science in the Federal Government*, p. 312.

[33]Willoughby, *Government Organization in War Time and After*, p. 25.

[34]Dupree, *Science in the Federal Government*, p. 312.

[35]See Zechariah Chafee, Jr. *Free Speech in the United States*. Cambridge, MA: Harvard University Press, 1941, part I; James R. Mock. *Censorship 1917*. Princeton, NJ: Princeton University Press, 1941; Paul L. Murphy. *World War I and the Origin of Civil Liberties in the United States*. New York: W.W. Norton and Company, 1979; Harry Scheiber. *The Wilson Administration and Civil Liberties, 1917–1921*. Ithaca, NY: Cornell University Press, 1960.

[36]Details on the organization and operation of the Committee may be found in George

Creel. *How We Advertised America*. New York: Harper and Brothers, 1920; James R. Mock and Cedric Larson. *Words that Won the War*. Princeton, NJ: Princeton University Press, 1939; U.S. Committee on Public Information. *Complete Report of the Chairman of the Committee on Public Information 1917: 1918: 1919*. Washington, DC: U.S. Govt. Print. Off., 1920; Stephen L. Vaughn. *Holding Fast the Inner Lines: Democracy, Nationalism, and the Committee on Public Information*. Chapel Hill: The University of North Carolina Press, 1980; Willoughby, *Government Organization in War Time and After*, pp. 33–38.

[37]Pursuant to 37 Stat. 302, 303; Susan J. Douglas. *Inventing American Broadcasting, 1899–1922*. Baltimore: The Johns Hopkins University Press, 1987, p. 276.

[38]40 Stat. 217, 219.

[39]40 Stat. 221.

[40]40 Stat. 225–226.

[41]40 Stat. 230–231.

[42]40 Stat. 411, 412–413.

[43]40 Stat. 425–426.

[44]40 Stat. 553–554.

[45]See, generally, Dorothy Ganfield Fowler. *Unmailable: Congress and the Post Office*. Athens: The University of Georgia Press, 1977, pp. 109–125.

[46]Dupree, *Science in the Federal Government*, pp. 312–313.

[47]Vaughn, *Holding Fast the Inner Lines*, p. 225.

[48]Mock, *Censorship 1917*, p. 135.

[49]*Ibid.*, pp. 137–138.

[50]55 Stat. 795.

[51]55 Stat. 796–797.

[52]55 Stat. 838.

[53]See, generally, Harwood L. Childs. "Public Information and Opinion." *American Political Science Review*, v. 37, February, 1943, pp. 56–68; Harwood L. Childs, ed. "The Office of War Information." *Public Opinion Quarterly*, v. 7, Spring, 1943; Elmer Davis and Byron Price. *War Information and Censorship*. Washington, DC: American Council on Public Affairs, 1943; Byron Price. "Governmental Censorship in War-time." *American Political Science Review*, v. 36, October, 1942, pp. 837–850; Robert E. Summers, comp. *Wartime Censorship of Press and Radio*. New York: H.W. Wilson Company, 1942.

[54]Anthony Cave Brown and Charles B. MacDonald, eds. *The Secret History of the Atomic Bomb*. New York: Dial Press/James Wade, 1977, pp. 30–31; Spencer R. Weart. "Scientists with a Secret." *Physics Today*, v. 29, February, 1976, pp. 23–30.

[55]James V. Forrestal. Speech before the Voluntary Security Conference, March 3, 1948, Washington, DC, appearing in Robert E. Summers, ed. *Federal Information Controls in Peacetime*. New York: H.W. Wilson Company, 1949, pp. 102–104.

[56]See *Ibid.*, pp. 104–109.

[57]James Russell Wiggins. *Freedom or Secrecy*. Revised edition. New York: Oxford University Press, 1964, p. 100.

[58]See *Federal Register*, v. 19, December 7, 1954, pp. 8045–8046; also see *Ibid.*, v. 20, September 28, 1955, p. 7233.

[59]U.S. Department of Commerce. Office of the Secretary. Press release (G-520). November 5, 1954. Washington, DC; also see Wiggins, *Freedom or Secrecy*, pp. 102–103.

[60]See U.S. Congress. House. Committee on Government Operations. *Availability of Information from Federal Departments and Agencies*. Hearings, 84th Congress, 2d Session. Washington, DC: U.S. Govt. Print. Off., 1956, pp. 1123–1187, 1233–1286, 1447–1521, 1639–1711.

[61]See U.S. Congress. House. Committee on Government Operations. *Availability of Information from Federal Departments and Agencies*. H. Rept. 2947, 84th Congress, 2d Session. Washington, DC: U.S. Govt. Print. Off., 1956, p. 91.

[62]See *Congressional Record*, v. 103, April 9, 1957, p. 5376; also see U.S. Congress. House. Committee on Government Operations. *Availability of Information from Federal Departments and Agencies*. H. Rept. 2578, 85th Congress, 2d Session. Washington, DC: U.S. Govt. Print. Off., 1958, pp. 13–14.

[63]See *Federal Register*, v. 22, July 24, 1957, p. 5876.

[64]See Deborah Shapley. "Intelligence Agency Chief Seeks 'Dialogue' With Academics." *Science*, v. 202, October 27, 1978, p. 407.

[65]"Report of the Public Cryptography Study Group." Prepared for the American Council on Education, Washington, D.C., February 7, 1981, p. 1; also appears in *Academe*, v. 67, December, 1981, p. 372.

[66]"Report of the Public Cryptography Study Group," p. 14; *Academe*, v. 67, December, 1981, p. 379.

[67]American Council on Education. Press release. April 20, 1981. Washington, DC, p. 3.

[68]See *Federal Register*, v. 47, March 1, 1982, p. 8704.

[69]George I. Davida. "The Case Against Restraints on Non-Governmental Research in Cryptography: A Minority Report of the Public Cryptography Study Group of the American Council on Education," Washington, D.C., February 7, 1981, p. 6; *Academe*, v. 67, December, 1981, p. 382.

[70]U.S. National Science Foundation. Mathematical and Computer Sciences Advisory Committee. *The Role of the NSF in Supporting Cryptological Research*. July 13, 1981. Washington, DC, p. 9.

[71]*Ibid.*, pp. 6–7; also see John Walsh. "Shunning Cryptocensorship." *Science*, v. 212, June 12, 1981, p. 1250.

[72]Jonathan Knight. "Cryptographic Research and NSA." *Academe*, v. 67, December, 1981, p. 371.

[73]*Ibid.*, p. 372.

[74]See Bobby R. Inman. "National Security and Technical Information" in Harold C. Relyea, ed. *Striking A Balance: National Security and Scientific Freedom—First Discussions*. Washington, DC: American Association for the Advancement of Science, 1985, pp. 51–55.

[75]U.S. Congress. House. Committee on Science and Technology. *Impact of National Security Considerations on Science and Technology*. Hearing, 97th Congress, 2d Session. Washington, DC: U.S. Govt. Print. Off., 1982, pp. 7–8, 41.

[76]*Ibid.*, pp. 7–8.

[77]*Ibid.*, p. 42.

[78]See 1 Stat. 28, 49, 65, 68, 553; 9 Stat. 395; 16 Stat. 163; 17 Stat. 283; and 18 Stat. 169. With the compilation of the *Revised Statutes of the United States* (1874), these separate authorities were combined in a single provision at section 161 which was continued in various editions of the *United States Code* without alteration until 1958 when a single amending sentence was added (72 Stat. 547); the provision presently is located at 5 U.S.C. 301.

[79]11 Stat. 60.

[80]See, generally, Harold C. Relyea. "The Presidency and the People's Right to Know" in Harold C. Relyea, ed. *The Presidency and Information Policy*. New York: Center for the Study of the Presidency, 1981, pp. 9–15.

[81]James L. McCamy. *Government Publicity: Its Practice in Federal Administration*. Chicago: University of Chicago Press, 1939, p. 103n.

[82]See 52 Stat. 3.

[83]See 3 C.F.R., 1938–1943 Comp., pp. 634–635.

[84]See Robert E. Summers, ed. *Federal Information Controls in Peacetime*, pp. 141–146.

[85]See 3 C.F.R., 1949–1953 Comp., pp. 298–299.

[86]See *Ibid.*, pp. 789–797.

[87]See 3 C.F.R., 1971–1975 Comp., pp. 678–690.

[88]See *Federal Register*, v. 43, July 3, 1978, pp. 28949–28962.

[89]U.S. Department of Defense. Office of the Director of Defense Research and Engineering. *Report of the Defense Science Board Task Force on Secrecy*. July 1, 1970. Washington, DC, p. iv.

[90]See *Federal Register*, v. 47, April 6, 1982, pp. 14874–14884; also see *Ibid.*, April 12, 1982, p. 15557; *Ibid.*, May 11, 1982, pp. 20105–20106; *Ibid.*, June 25, 1982, pp. 27836–27842.

[91]40 Stat. 394.

[92]40 Stat. 411, 422.

[93]54 Stat. 710.

[94]See 56 Stat. 370.

[95]66 Stat. 3; 66 Stat. 792, 805.

[96]See 50 U.S.C. App. 10(i).

[97]35 U.S.C. 181.

[98]35 U.S.C. 184.

[99]35 U.S.C. 186.

[100]35 U.S.C. 182.

[101]See U.S. Congress. House. Committee on Government Operations. *The Government's Classification of Private Ideas*. Hearings, 96th Congress, 2d Session. Washington, DC: U.S. Govt. Print. Off., 1980, pp. 241–290.

[102]35 U.S.C. 183.

[103]See Mary M. Cheh. "The *Progressive* Case and the Atomic Energy Act: Waking to the Dangers of Government Information Controls." *George Washington Law Review*, v. 48, January, 1980, p. 167n, citing *Halpern* v. *United States*, 258 F.2d 36, 38–39 (2d Cir. 1958).

[104]Edmund L. Andrews. "Cold War Secrecy Still Shrouds Inventions." *New York Times*, May 23, 1992, p. 35.

[105]40 Stat. 411.

[106]54 Stat. 712, 714.

[107]See 56 Stat. 463; 58 Stat. 671; 59 Stat. 270; 60 Stat. 215; 61 Stat. 214, 323, 946.

[108]63 Stat. 7.

[109]William J. Long. *U.S. Export Control Policy*. New York: Columbia University Press, 1989, p. 23.

[110]See U.S. Congress. House. Committee on Government Operations. *The Government's Classification of Private Ideas*. H. Rept. 96–1540, 96th Congress, 2d Session. Washington, DC: U.S. Govt. Print. Off., 1980, pp. 98–99.

[111]President's Scientific Research Board. *Science and Public Policy: Administration for Research*. Vol. 3. Washington, DC: U.S. Govt. Print. Off., 1947, p. 37.

[112]U.S. Department of State. *Science and Foreign Relations*. Washington, DC: Department of State, 1950, p. 85.

[113]See U.S. Congress. House. Committee on Government Operations. *Availability of Information from Federal Executive Agencies*. Hearings, 84th Congress, 2d Session. Washington, DC: U.S. Govt. Print. Off., 1956, pp. 725–922, 1123–1711.

[114]See U.S. Congress. House. Committee on Government Operations. *Availability of Information from Federal Departments and Agencies*. H. Rept. 2947, 84th Congress, 2d Session. Washington, DC: U.S. Govt. Print. Off., 1956, p. 92; also see U.S. Congress. House. Committee on Government Operations. *Availability of Information from Federal Departments and Agencies*. H. Rept. 2578, 85th Congress, 2d Session. Washington, DC: U.S. Govt. Print. Off., 1958, pp. 17–28.

[115]83 Stat. 841.

[116]See Long, *U.S. Export Control Policy*, pp. 29–53; Stuart Macdonald. *Technology and the Tyranny of Export Controls*. Houndmills: The Macmillan Press, 1990, pp. 20–21.

[117]See *Ibid.*, pp. 55–67.

[118]93 Stat. 503.

[119]U.S. Department of Defense. Office of the Director of Defense Research and Engineering. *An Analysis of Export Control of U.S. Technology—A DOD Perspective: A Report*

of the Defense Science Board Task Force on Export of U.S. Technology. February 4, 1976. Washington, DC. pp. 27–30.

[120]See *Federal Register*, v. 45, October 1, 1980, pp. 65014–65019; the Department of Energy, in support of the Department of Defense, prepared a list of energy-related militarily critical technologies, but only the table of contents was published in *Ibid*., pp. 65152–65167.

[121]See Long, *U.S. Export Control Policy*, pp. 69–89.

[122]However, as early as 1951, Congress had approved, in the Mutual Defense Assistance Control Act, better known as the Battle Act, diplomatic efforts on the part of the United States at establishing with its allies and other countries "a program for controlling exports . . . to any nation or combination of nations threatening the security of the United States, including the Union of Soviet Socialist Republics and all countries under its domination." See 65 Stat. 644, 646–647. The statute denied U.S. aid for economic recovery to Western European countries not joining the COCOM system. Stuart Macdonald. *Technology and the Tyranny of Export Controls*, pp. 17, 25.

[123]Stuart Macdonald. *Technology and the Tyranny of Export Controls*, pp. 15–17.

[124]*Ibid*., pp. 16, 117–118.

[125]U.S. Congress. House. Committee on Foreign Affairs. *Technology Exports: Department of Defense Organization and Performance*. Hearing, 96th Congress, 1st Session. Washington, DC: U.S. Govt. Print. Off., 1980.

[126]99 Stat. 120.

[127]See Long, *U.S. Export Control Policy*, pp. 91–99.

[128]See U.S. Congress. Senate. Committee on Banking, Housing, and Urban Affairs. *Export Administration Act*. Hearing, 98th Congress, 1st Session. Washington, DC: U.S. Govt. Print. Off., 1983; U.S. Congress. Senate. Committee on Banking, Housing, and Urban Affairs. *Reauthorization of the Export Administration Act*. Hearings, 98th Congress, 1st Session. Washington, DC: U.S. Govt. Print. Off., 1983; U.S. Congress. House. Committee on Foreign Affairs. *Extension and Revision of the Export Administration Act of 1979*. Hearings and mark-up, 98th Congress, 1st Session. Washington, DC: U.S. Govt. Print. Off., 1985; U.S. Congress. House. Committee on Armed Services. *Technology Transfer*. Hearings, 98th Congress, 1st Session. Washington, DC: U.S. Govt. Print. Off., 1984; U.S. Congress. House. Committee on Armed Services. *Full Committee Consideration of H.R. 3231 to Amend the Authorities Contained in the Export Administration Act of 1979, and for Other Purposes*. Hearing, 98th Congress, 1st Session. Washington, DC: U.S. Govt. Print. Off., 1984.

[129]See, for example, U.S. Congress. Senate. Committee on Governmental Affairs. *Transfer of United States High Technology to the Soviet Union and Soviet Bloc Nations*. Hearings, 97th Congress, 2d Session. Washington, DC: U.S. Govt. Print. Off., 1982; U.S. Congress. Senate. Committee on Governmental Affairs. *Transfer of Technology*. Hearings, 98th Congress, 2d Session. Washington, DC: U.S. Govt. Print. Off., 1984; U.S. Congress. House. Committee on Science and Technology. *Impact of National Security Considerations on Science and Technology*. Hearing, 97th Congress, 2d Session. Washington, DC: U.S. Govt. Print. Off., 1982; U.S. Congress. House. Committee on Science and Technology. *Scientific Communications and National Security*. Hearing, 98th Congress, 2d Session. Washington, DC: U.S. Govt. Print. Off., 1984; U.S. Congress. House. Committee on Science and Technology. *Technology Transfer*. Hearings, 99th Congress, 1st Session. Washington, DC: U.S. Govt. Print. Off., 1985; U.S. Congress. House. Committee on the Judiciary. *1984: Civil Liberties and the National Security State*. Hearings, 98th Congress, 1st and 2d Sessions. Washington, DC: U.S. Govt. Print. Off., 1984.

[130]U.S. Congress. House. Committee on Foreign Affairs. *Extension and Revision of the Export Administration Act of 1979*, pp. 1259–1263.

[131]U.S. Congress. House. Committee on Foreign Affairs. *Export Administration Act Amendments of 1983*. H. Rept. 98–257 Part 1, 98th Congress, 1st Session. Washington, DC: U.S. Govt. Print. Off., 1983, p. 16.

[132]99 Stat. 120, 121.

[133]The language comes largely from a Senate provision; see U.S. Congress. Senate. Committee on Banking, Housing, and Urban Affairs. *The Export Administration Act Amendments of 1983*. S. Rept. 98–170, 98th Congress, 1st Session. Washington, DC: U.S. Govt. Print. Off., 1983, pp. 25, 38.

[134]99 Stat. 120, 155.

[135]102 Stat. 1107, 1362.

[136]See U.S. Congress. House. Committee on Foreign Affairs. *The Reauthorization of the Export Administration Act*. Hearings and markup, 101st Congress, 2d Session. Washington, DC: U.S. Govt. Print. Off., 1990; U.S. Congress. Senate. Committee on Banking, Housing, and Urban Affairs. *Reauthorization of the Export Administration Act*. Hearings, 101st Congress, 2d Session. Washington, DC: U.S. Govt. Print. Off., 1990.

[137]See *Weekly Compilation of Presidential Documents*, v. 26, November 19, 1990, pp. 1839–1840.

[138]See E.O. 12730, *Federal Register*, v. 55, October 2, 1990, pp. 40373–40374.

[139]See the text accompanying notes 94–98 of chapter 2.

[140]22 U.S.C. 2778; see 22 U.S.C. 2751–2794.

[141]68 Stat. 832.

[142]75 Stat. 424.

[143]82 Stat. 1320.

[144]22 C.F.R. Part 120–130 (1989).

[145]22 C.F.R. 120.10.

[146]22 C.F.R. 120.21.

[147]22 C.F.R. 120.18.

[148]See 22 C.F.R. Part 121.

[149]22 U.S.C. 2778(e).

[150]22 U.S.C. 2778(c).

[151]See Richard G. Hewlett and Oscar E. Anderson, Jr. *A History of the United States Atomic Energy Commission: The New World, 1939/1946*. Vol. 1. University Park: Pennsylvania State University Press, 1962, pp. 14–71.

[152]For an overview of the development and operations of the Manhattan Engineer District, see *Ibid*., pp. 71–407, 620–655; Brown and MacDonald, *The Secret History of the Atomic Bomb*; Leslie R. Groves. *Now It Can Be Told: The Story of the Manhattan Project*. New York: Harper and Brothers, 1962.

[153]Groves, *Now It Can Be Told*, pp. 138–139.

[154]*Ibid*., p. 140.

[155]Brown and MacDonald, *The Secret History of the Atomic Bomb*, p. 204.

[156]*Ibid*.

[157]*Ibid*.

[158]60 Stat. 755.

[159]68 Stat. 919; codified at 42 U.S.C. 2011 et. seq. Concerning the 1954 modifications and their implementation, see, generally, Richard G. Hewlett and Jack M. Holl. *Atoms For Peace and War, 1953–1961*. Berkeley: University of California Press, 1989.

[160]42 U.S.C. 2014(y).

[161]See Richard G. Hewlett. "The 'Born-Classified' Concept in the U.S. Atomic Energy Commission" in U.S. Congress. House. Committee on Government Operations. *The Government's Classification of Private Ideas*. H. Rept. 96–1540, 96th Congress, 2d Session. Washington, DC: U.S. Govt. Print. Off., 1980, pp. 173–180.

[162]See *United States* v. *The Progressive, Inc.*, 467 F.Supp. 990 (E.D. Wis. 1979); also see Mary M. Cheh. "The *Progressive* Case and the Atomic Energy Act: Waking to the Dangers of Government Information Controls." *George Washington Law Review*, v. 48, January, 1980, pp. 163–210; Harold C. Relyea. "Information, Secrecy, and Atomic Energy." *New York University Review of Law and Social Change*, v. 10, no. 2, 1980–1981, pp. 265–286.

[163]U.S. Office of Technology Assessment. *Science, Technology, and the First Amendment*. Washington, DC: U.S. Govt. Print. Off., 1988, p. 55.

[164]*Ibid*., p. 57.

[165]97 Stat. 614, 690; initially codified at 10 U.S.C. 140c (1983) and subsequently renumbered as 10 U.S.C. 130 (1988).

[166]See U.S. Department of Defense. Directive 5230.24. "Distribution Statements on Technical Documents." Washington, DC, November 20, 1984; U.S. Department of Defense. Directive 5230.25. "Withholding of Unclassified Technical Data From Public Disclosure." Washington, DC, November 6, 1984; 32 C.F.R. Part 250 (1990).

[167]See *Federal Register*, v. 51, September 17, 1986, pp. 32938–32940.

[168]See, generally, Harold C. Relyea. "The Coming of Secret Law." *Government Information Quarterly*, v. 5, no. 2, 1988, pp. 97–116.

[169]For the text of NSDD 84, see U.S. Congress. Senate. Committee on Governmental Affairs. *National Security Decision Directive 84*. Hearing, 98th Congress, 1st Session. Washington, DC: U.S. Govt. Print. Off., 1984, pp. 85–86.

[170]See U.S. Congress. House. Permanent Select Committee on Intelligence. *Security Clearance Procedures in the Intelligence Agencies: Staff Report*. Committee print, 96th Congress, 1st Session. Washington, DC: U.S. Govt. Print. Off., 1979; U.S. Congress. House. Permanent Select Committee on Intelligence. *Prepublication Review and Secrecy Agreements*. Hearings, 96th Congress, 2d Session. Washington, DC: U.S. Govt. Print. Off., 1980; U.S. Congress. House. Committee on the Judiciary. *Prepublication Review and Secrecy Requirements Imposed Upon Federal Employees*. Hearings, 96th Congress, 2d Session. Washington, DC: U.S. Govt. Print. Off., 1980.

[171]*Snepp* v. *United States*, 444 U.S. 507 (1980).

[172]See U.S. Office of Technology Assessment. *Scientific Validity of Polygraph Testing: A Research Review and Evaluation*. Washington, DC: U.S. Govt. Print. Off., 1983.

[173]See, for example, U.S. Congress. Senate. Committee on Governmental Affairs. *National Security Decision Directive 84*. Hearing, 98th Congress, 1st Session. Washington, DC: U.S. Govt. Print. Off., 1984; U.S. Congress. House. Committee on Government Operations. *Review of the President's National Security Decision Directive 84 and the Proposed Department of Defense Directive on Polygraph Use*. Hearing, 98th Congress, 1st Session. Washington, DC: U.S. Govt. Print. Off., 1984; U.S. Congress. House. Committee on Government Operations. *The Administration's Initiatives to Expand Polygraph Use and Impose Lifelong Censorship on Thousands of Government Employees*. H. Rept. 98–578, 98th Congress, 1st Session. Washington, DC: U.S. Govt. Print. Off., 1984; U.S. Congress. House. Committee on the Judiciary. *Presidential Directive on the Use of Polygraphs and Prepublication Review*. Hearings, 98th Congress, 1st and 2d Sessions. Washington, DC: U.S. Govt. Print. Off., 1985.

The persistence of the issue prompted additional hearings into the latter half of the decade: U.S. Congress. House. Committee on Post Office and Civil Service and Committee on the Judiciary. *Protection of Classified Information*. Joint hearing, 99th Congress, 1st Session. Washington, DC: U.S. Govt. Print. Off., 1987; U.S. Congress. House. Committee on Post Office and Civil Service. *Classified Information Nondisclosure Agreements*. Hearing, 100th Congress, 1st Session. Washington. DC: U.S. Govt. Print. Off., 1988; U.S. Congress. House. Committee on Government Operations. *Congress and the Administration's Secrecy Pledges*. Hearing, 100th Congress, 2d Session. Washington, DC: U.S. Govt. Print. Off., 1988.

[174]*Congressional Record*, v. 129, July 15, 1983, pp. 19479–19482.

[175]*Ibid*., July 26, 1983, pp. 20977–20978.

[176]Private sector use of polygraph testing was later restrained in 1988 with the enactment of the Employee Polygraph Protection Act, 102 Stat. 646.

[177]97 Stat. 614–615.

[178]*Congressional Record*, v. 129, October 20, 1983, pp. 28702–28720.

[179]97 Stat. 1061.

[180]H.R. 4681, *Congressional Record*, v. 130, January 3, 1984, p. E178.

[181]See U.S. Congress. House. Committee on Post Office and Civil Service. *Federal Polygraph Limitation and Anti-Censorship Act of 1984*. Hearing, 98th Congress, 1st Session. Washington, DC: U.S. Govt. Print. Off., 1984; U. S. Congress. House. Committee on Post Office and Civil Service. *Federal Polygraph Limitation and Anti-Censorship Act of 1984*. H. Rept. 98–961 Part 1, 98th Congress, 1st Session. Washington, DC: U.S. Govt. Print. Off., 1984; U.S. Congress. House. Committee on Armed Services. *Hearings on H.R. 4681*. Hearing, 98th Congress, 2d Session. Washington, DC: U.S. Govt. Print. Off., 1985; U.S. Congress. House. Committee on Armed Services. *Full Committee Consideration of H.R. 4681*. Hearing, 98th Congress, 2d Session. Washington, DC: U.S. Govt. Print. Off., 1984; U.S. Congress. House. Committee on Armed Services. *Federal Polygraph Limitation and Anti-Censorship Act of 1984*. H. Rept. 98–961 Part 2, 98th Congress, 2d Session., Washington, DC: U.S. Govt. Print. Off., 1984; U.S. Congress. House. Committee on the Judiciary. *Federal Polygraph Limitation and Anti-Censorship Act*. Hearing, 98th Congress, 2d Session. Washington, DC: U.S. Govt. Print. Off., 1986.

[182]See Leslie Maitland Werner. "Reagan to Relent on Secrecy Pledge." *New York Times*, February 15, 1984, p. A20; David Hoffman. "Reagan Shelves Requirement for Polygraph Tests." *Washington Post*, February 16, 1984, pp. A1, A12; Stuart Taylor, Jr. "Secrecy Proposal: Risks Weighed Against Gains." *New York Times*, February 20, 1984, p. A13.

[183]See George Lardner, Jr. "McFarlane Acts on Reagan's Orders: New Polygraphs, Censorship Shelved." *Washington Post*, March 22, 1984, p. A21; Stuart Taylor, Jr. "Delay Continued on Security Rule: White House Retreats Again From Its Censorship Plan." *New York Times*, March 24, 1984, p. 8.

[184]See U.S. General Accounting Office. *Information and Personnel Security: Data on Employees Affected by Federal Security Programs*. GAO/NSIAD-86-189FS. Washington, DC: U.S. Govt. Print. Off., 1986; also see U.S. Congress. House. Committee on Post Office and Civil Service. *Classified Information Nondisclosure Agreements*. Hearing, 100th Congress, 1st Session. Washington, DC: U.S. Govt. Print. Off., 1988.

[185]101 Stat. 1329–432; continued by 102 Stat. 1756 and 103 Stat. 820.

[186]See *National Federation of Federal Employees* v. *United States*, 688 F. Supp. 671 (D.D.C. 1988) and 695 F. Supp. 1196 (D.D.C. 1988).

[187]See U.S. Congress. House. Committee on Government Operations. *Congress and the Administration's Secrecy Pledges*. Hearing, 100th Congress, 2d Session. Washington, DC: U.S. Govt. Print. Off., 1989.

[188]See *American Foreign Service Association* v. *Garfinkel*, 490 U.S. 153 (1989).

[189]104 Stat. 1475.

[190]U.S. Congress. House. Committee of Conference. *Making Appropriations for the Treasury Department, the U.S. Postal Service, the Executive Office of the President, and Certain Independent Agencies, for the Fiscal Year Ending September 30, 1991, and for Other Purposes*. H. Rept. 101–906, 101st Congress, 2d Session. Washington, DC: U.S. Govt. Print. Off., 1990, p. 96.

[191]See *Federal Register*, v. 56, January 23, 1991, pp. 2644–2645.

[192]Compiling the results of a survey conducted by Representative John Conyers, Jr., chairman of the House Subcommittee on Legislation and National Security, the General Accounting Office reported that of 54 federal agencies receiving the Subcommittee's questionnaire, "48 responded that government and contractor employees signed about 144,000 and 98,000 nondisclosure agreements, respectively, between October 1, 1989, and March 31, 1990. According to the agencies, their cost to implement the signing of these agreements during this 6-month period were about $433,000 for government employees and about $12,000 for contractor employees." These same 48 agencies revealed that government and contractor employees signed about 338,000 and 1,279,000 nondisclosure

agreements, respectively, during 1988. They also reported that government employees signed about 241,000 agreements and contractors signed about 176,000 agreements in 1989. U.S. General Accounting Office. *Federal Agency Use of Nondisclosure Agreements.* GAO/NSIAD-91-106FS. Washington, DC: U.S. Govt. Print. Off., 1991, p. 2.

[193]Harold P. Green. "The Boundaries of Scientific Freedom." *Newsletter on Science, Technology, and Human Values,* June, 1977, p. 17.

[194]Steven Goldberg. "The Constitutional Status of American Science." *University of Illinois Law Forum,* v. 1979, no. 1, 1979, p. 14.

[195]James R. Ferguson. "Scientific Inquiry and the First Amendment." *Cornell Law Review,* v. 64, April, 1979, p. 644; also see *Federal Communications Commission* v. *Pacifica Foundation,* 438 U.S. 726, 746 (1978); *Miller* v. *California,* 413 U.S. 15, 34 (1973); *Roth* v. *United States,* 354 U.S. 476, 484 (1957).

[196]Ferguson, "Scientific Inquiry and the First Amendment," p. 645.

[197]*Ibid.,* pp. 645–648; also see James R. Ferguson. "Scientific and Technological Expression: A Problem in First Amendment Theory." *Harvard Civil Rights-Civil Liberties Law Review,* v. 16, Fall, 1981, pp. 519–560; James R. Ferguson. "Scientific Freedom, National Security, and the First Amendment." *Science,* v. 221, August 12, 1983, pp. 620–624.

[198]*Virginia State Board of Pharmacy* v. *Virginia Citizens Consumer Council, Inc.,* 425 U.S. 748, 761–765 (1976).

[199]Ferguson, "Scientific Inquiry and the First Amendment," p. 645.

[200]*Ibid.,* p. 648.

[201]Ferguson, "Scientific Freedom, National Security, and the First Amendment, p. 623.

[202]80 Stat. 250; codified at 5 U.S.C. 552.

[203]*Environmental Protection Agency* v. *Mink,* 410 U.S. 73 (1973).

[204]88 Stat. 1561.

[205]See Ronald M. Levin. "In Camera Inspections Under the Freedom of Information Act." *University of Chicago Law Review,* v. 41, Spring, 1974, pp. 557–581; Note. "National Security and the Public's Right to Know: A New Role for the Courts Under the Freedom of Information Act." *University of Pennsylvania Law Review,* v. 123, June, 1975, pp. 1438–1473; Note. "National Security and the Amended Freedom of Information Act." *Yale Law Journal,* v. 85, January, 1976, pp. 401–422.

[206]See *Weissman* v. *Central Intelligence Agency,* 565 F.2d 692 (D.C. Cir. 1977); also see Robert W. Coykendall. "*In Camera* Inspection of National Security Files Under the Freedom of Information Act." *Kansas Law Review,* v. 26, Summer, 1978, pp. 617–624.

[207]*New York Times Company* v. *United States,* 403 U.S. 714 (1971).

[208]See *Ibid.,* p. 718; also see *Ibid.,* pp. 720–721 (Justice Douglas concurring), p. 732 (Justice White concurring), and pp. 742–747 (Justice Marshall concurring).

[209]*Ibid.,* p. 730.

[210]*Ibid.,* pp. 730, 731. Interestingly, Erwin N. Griswold, who, as Solicitor General, argued the government's case for restraining press publication of the Pentagon Papers before the Supreme Court, subsequently conceded he has "never seen any trace of a threat to the national security from the publication." See Erwin N. Griswold. "Secrets Not Worth Keeping." *Washington Post,* February 15, 1989, p. A25; Saundra Torry. "At 87, Erwin N. Griswold Is the Dean of Supreme Court Observers." *Washington Post Washington Business,* July 15, 1991, p. 5.

[211]See *United States* v. *Progressive, Inc.,* 467 F. Supp. 990 (E.D. Wis. 1979).

[212]*Ibid.*

[213]See *United States* v. *Edler Industries, Inc.,* 579 F.2d 516 (9th Cir. 1978); in *United States* v. *Swaroski,* 592 F.2d 1341 (2nd Cir. 1979), the appellant was convicted for attempting to export a military aircraft gunsight camera without a license, and was unsuccessful in contending that the International Traffic in Arms licensing provisions were unconstitu-

tionally vague. Also see *United States* v. *Gurrola-Garcia,* 547 F.2d 1975 (9th Cir. 1976); *United States* v. *Stone,* 452 F.2d 42 (8th Cir. 1971); *Samora* v. *United States,* 406 F.2d 1095 (5th Cir. 1969).

[214]See *Zemel* v. *Rusk,* 381 U.S. 1 (1965); *Kent* v. *Dulles,* 357 U.S. 116 (1958).

[215]*Haig* v. *Agee,* 453 U.S. 291, 307 (1981).

[216]*Snepp* v. *United States,* 444 U.S. 507, 509 n. 3 (1980).

4

THE CAMPUS SETTING

In the aftermath of World War II, significant new developments occurred in federal financing of scientific research. Among them was the funding of campus-based scientists for the performance of basic research, setting a pattern that has persisted to the moment. Indeed, the universities quickly became, and have remained, a valuable and fruitful source of basic research. Coupled with this development was the continued rise in federal expenditures for scientific research. Also, new institutional patterns emerged. After a 5-year struggle, a central scientific agency, the National Science Foundation, was established within the federal government in 1950.[1] Furthermore, although this entity became an important source of funding for campus-based scientific research, another governmental financier also soon made its presence felt.

The Department of Defense began to emerge in 1947 with the creation of the National Military Establishment,[2] and attained structural maturity 2 years later.[3] The existing War (renamed Army) and Navy departments, along with a new Air Force department, became DOD subunits subject to the supervision of a Secretary of Defense who was the chief manager of and sole Cabinet representative for the armed services agencies.

CAMPUS RESEARCH

Prior to World War II, military support of scientific research at colleges and universities was essentially nonexistent. During the war, however, campus-based scientists and engineers performed research for the armed services with funding from the Office of Scientific Research and Development, a presidential agency directed by Vannevar Bush.[4] After the war, the Navy Department was the first among the military entities to structure itself to facilitate university research by establishing the Office of Naval Research in 1946. The Army Department followed this example in 1951 with the creation of its Office of Ordinance Research, reconstituted as the Army Research Office in 1961. The Air Force Department instituted its Office of Scientific Research in 1951 as well. The first of these units, ONR, allocated substantial funding for campus-based scientific research and, by 1950, was supporting over 40% of U.S. basic science.[5]

DOD largely came to dominate federal research and development support from the early 1950s into the mid-1960s. Growing campus opposition to the Vietnam War, combined with more active research sponsorship by the National Science Foundation, the National Institutes of Health, and the National Aeronautics and Space Administration, subsequently began to reduce DOD's commanding role. By 1975, universities were receiving only 8% of all of their federal research and development funds from the Pentagon.[6] A change of administration and a new Secretary of Defense, Harold Brown, started reversing this trend, which was accelerated by the Reagan regime's defense chief, Caspar W. Weinberger. At the close of the Reagan presidency, defense-related research and development obligations—funded by DOD and the Department of Energy—constituted over 70% of all federal monies allocated for this purpose.

A similar pattern prevailed regarding DOD support of campus-based research and development activities. In the late 1950s and early 1960s, the Pentagon dominated federal funding, sponsoring in 1958, for example, 44% of the total federal research and development allocation and 41% of all the basic research performed at U.S. universities. The Department's support for such basic research grew from $56 million in 1958 to $132 million in 1965, an increase of 136%. Overall, DOD university support for research and development jumped from $118 million in 1958 to $291 million in 1965, a 145% hike.[7]

In the mid-1960s, students and faculty on many campuses became engaged in active and vigorous protest of U.S. involvement in the Vietnam War. A distaste for the military establishment began to grow within the nation's colleges and universities. It was fueled by increasing disapproval of American operations in Southeast Asia, objections to the military draft, displeasure with secret research being performed for the Pentagon by universities, and a perception that any DOD presence on campus or in academic life was, for the most part, inconsistent with the mission and functions of a university.

This climate of opinion prompted university and college administrators and faculties to take various actions to limit the conduct of both classified and unclassified DOD-sponsored research on campus. For example, the Faculty Senate of the State University of New York at Stony Brook passed a resolution in May 1970, declaring "that it be SUNY/SB policy not to seek research grants and contracts from the Department of Defense."[8] That same month, at Princeton University, faculty and students adopted two resolutions, one urging Congress "to seek methods of transferring funds for basic research from the Department of Defense to civilian agencies as soon as possible" and another stating that the chemistry department "will henceforth discourage all faculty members from accepting any new grants from DOD or any of its subagencies." Faculty conducting studies funded by DOD "are requested to seek other sources of support as soon as possible," said the resolution. During the year, both the Massachusetts Institute of Technology and Stanford University announced they were dropping their association with DOD-supported research centers—respectively, the MIT Draper Laboratory and the Stanford Research Institute. Subsequently, DOD withdrew support from 16 such federally funded research and development centers associated with several different universities.[9]

Due to a combination of factors, not the least of which were budget restraints resulting from the Vietnam conflict, as well as campus opposition to Pentagon-sponsored studies performed by the academic community, DOD funding for research and development at the nation's universities dropped from $291 million in 1965 to $203 million in 1975, a 31% decrease.[10] The situation improved for DOD with the conclusion of the Vietnam War. Important new leadership arrived in the person of Harold Brown as Secretary of Defense. A trained physicist and veteran of various federal science positions, including Director of Defense Research and Engineering and Secretary of the Air Force, Brown knew the value of the campus science research relationship for DOD and was eager to renew it. Furthermore, he was encouraged by a 1978 report prepared by an Office of Science and Technology Policy expert advisory group led by J. K. Galt of Sandia Laboratories. The Galt report offered a number of strong recommendations calling for an increase in DOD support for basic research, saying it was essential for the modernization of U.S. military forces, for direct access by DOD to available scientific and engineering talent that could be employed to help solve future scientific and technological challenges, and for acquainting a diverse group of scientists and engineers with the Department's research environment and its particular research needs. In addition, a broad approach to research sponsorship was encouraged. "In deciding which proposed projects to support," said the report, "the research manager in government should not use narrowly-identified relevance as a filter through which every project should pass, but rather should apply a broad concept to wide areas of science."[11]

Brown's successor, Caspar Weinberger, in seeking to close the nation's so-called "window of vulnerability" and to maintain, if not advance, the scientific and technological superiority of the U.S. defense position, further accelerated basic research funding. Issuing a memorandum in the summer of 1981 calling the current national base of support for basic research inadequate to meet future DOD needs, Weinberger eventually doubled Pentagon funds for basic research at colleges and universities during his tenure as Secretary of Defense.[12]

By 1981, the first year of the Reagan Administration, many colleges and universities had experienced about three decades of variously performing scientific research and development under Pentagon sponsorship. Past student and faculty objections to some kinds of DOD projects, such as weapons development, had largely been met through a number of different arrangements. Some schools had decided that, as a matter of policy, they would not undertake security classified studies; others determined that their basic charters precluded any type of secret research. In some cases, off-campus institutions had been created to enable faculty scientists to engage in DOD-funded research and development activities, which sometimes required special security arrangements, without violating university rules. Furthermore, Pentagon research contracts were very clear and explicit regarding security requirements. In brief, scientists and engineers conducting militarily significant studies under federal sponsorship were well aware of the necessities of safeguarding and protecting their work. Moreover, unless research, whether governmentally or privately funded, was specifically required to be protected, the prevailing presumption within the U.S. campus science community was that it was open to observation, discussion, and participation by students and professional peers, at the discretion of those principally responsible for its performance.

NEW CONTROLS

As détente collapsed in 1979 and U.S. relations with the Soviet Union rigidified and became more tense, American university scientists discovered that some areas of scientific research and knowledge were subject to national security controls apart from security classification or DOD contract restrictions. Among the first academic affairs to be affected were research presentations at professional conferences, then the campus activities of some visiting foreign scholars, and, eventually, the study programs and courses available to students from certain countries.

The first such incidents occurred in February 1980, shortly after the Soviet invasion of Afghanistan. Officials from the Department of Commerce, acting only a few days before the meeting was to occur, informed American Vacuum Society organizers of a small international gathering on magnetic bubble memory devices that the presence of certain foreign nationals at the conference subjected the scheduled proceedings to compliance with export requirements. In order to avoid possible fines or imprisonment for failure to comply with export regulations, the sponsors of the meeting rescinded invitations that had been sent to attendees coming from Hungary, Poland, and the Soviet Union. Scientists from the People's Republic of China, who were already en route, were subsequently permitted to attend the conference, provided they signed an agreement not to "reexport" what they learned to any nationals from 18 named countries.[13] A Commerce Department official later indicated that this case marked the first time that the authority of the Export Administration Act had been invoked to apply to information conveyed at a scientific meeting.[14] It would not be the last such application.

A few days after this incident, the Department of State prohibited nine Soviet citizens from traveling to a San Diego conference on lasers, electro-optical systems, and inertial confinement fusion that had been jointly organized by the Institute of Electrical and Electronics Engineers and the Optical Society of America. Reasons given for the restriction reportedly were that certain sensitive technical equipment having military potential would be on open display at the gathering and that the United States was protesting the Soviet invasion of Afghanistan.[15]

In early 1981, the Department of State informed Cornell University that a Hungarian engineer, scheduled to visit the campus to study electronics, would have to be limited to classroom pursuits, could not participate in any private seminars or discussions, and would be prohibited from receiving prepublication copies of research papers. In view of these conditions, the visit was canceled. Elsewhere, at the Massachusetts Institute of Technology, a $250,000 Air Force contract for research on computer-aided design was found to contain a clause establishing a two-year federal control over the release of research results in order to deny access to foreign nationals. Finding the caveat objectionable, MIT declined the contract.[16]

By the spring of 1981, a minor tempest was brewing on several campuses over efforts by the Department of Defense to utilize export regulations to restrict the participation of foreign nationals in the Pentagon's Very High Speed Integrated Circuit (VHSIC) research program. VHSIC studies, performed by both universities and industries, were centrally concerned with circuit design and materials science. The effort was to develop microcircuits with very fast signal processing speeds and the ability, as well, to withstand intense heat and radiation. Such circuitry would be used to create the computerized intelligence of sophisticated weapons.

Responding to a congressional mandate making VHSIC research subject to export control under the International Traffic in Arms Regulations (ITAR), DOD research managers sent a memorandum in mid-December to VHSIC program directors offering guidance on the application of the ITAR. The guidelines called for a distinction between basic and applied research, the export regulations applying only to the latter area. In cases where there was some doubt as to whether research was of one type or the other, the matter was to be referred to DOD for a decision. Thus, VHSIC applied research could not be presented in open forums—conferences, symposia, and classrooms—or otherwise disclosed to individuals who were not U.S. citizens. Furthermore, the memorandum concluded, saying:

> In the case of basic research supported by the VHSIC program, although such research and its results are not generally controlled, it is the preference of the Program Office that only U.S. citizens and immigrant aliens who have declared their intention of becoming citizens participate. Where this preference cannot be accommodated, the contractor should be directed to the Program Office for resolution.[17]

In response, the presidents of Cornell University, MIT, the California Institute of Technology, the University of California, and Stanford University sent a jointly signed protest letter in late February 1981 to the Secretaries of Commerce, State, and Defense, collectively, "to request clarification of the applicability of certain export restrictions to teaching and research activities conducted by American universities" and indicating that they were "deeply concerned about recent attempts to apply to universities the International Traffic in Arms Regulations (ITAR) and Export Administration Regulations (EAR)." In concluding their letter, the university presidents expressed the "hope that after examining this issue carefully, you will clarify what has always been our understanding—namely, that the regulations are not intended to limit academic exchanges arising from unclassified research and teaching." The controversy, however, continued without formal resolve.[18]

Later that autumn, the Department of State began asking academic officials at various major universities to cooperate in prohibiting visiting students from the People's Republic of China—numbering about 2,000 annually—from engaging in research and studies in certain scientific and technological areas that were regarded as subject to export restrictions. Many campus administrators declined to engage in such policing actions. For example, C. Peter Magrath, president of the University of Minnesota, whose institution counted about 130 Chinese students, wrote the State Department: "Our mission is teaching, research and public service, and neither our faculty nor our administrators were hired to implement government security actions." In his letter, Magrath noted that Minnesota did no classified research. "Both in principle and in practice," he wrote, "the restrictions proposed in your letter are inappropriate for an American research university."[19]

During the spring of 1982, customs inspectors, acting under the authority of Federal export control regulations, halted the shipment of some computer science books that a professional society was sending to Japan. Although the books reportedly were available to the public in the United States, customs officials indicated they were being held pending detailed explanations of their contents and destination.[20]

Another incident occurred in May. Five Chinese scholars and students, while in New York, awaiting the takeoff of a flight returning them to their own country, were informed that federal customs agents were confiscating suspicious-looking items from their luggage. The objects in question included articles from scientific journals, classroom notebooks, thesis and lecture materials, slides, computer software, and tape recordings. The students had been studying engineering at the University of Michigan and Michigan State University. Supervising faculty at the two schools later indicated that none of the students was studying anything classified, directly related to military technology, or threatening to the security of the United States. All of the material was subsequently returned. The search was conducted as part of a new enforcement program—identified as *Operation Exodus* by a Commerce Department official—to check the outflow of high technology knowledge. Federal Bureau of Investigation agents also reportedly were involved in the luggage search and earlier had made inquiries at the University of Michigan concerning the books used by the visiting Chinese students.[21]

Around this same time, nationals from Communist bloc countries were denied online access to the National Library of Medicine's MEDLINE service, a computerized index of articles appearing in some 3,000 medical and biomedical journals of considerable value to students, researchers, and practitioners. The action reportedly was prompted by concerns within the Department of Commerce that a skillful computer operator might jump from MEDLINE to some other "sensitive" file. Without online access, Communist bloc nationals had to make their MEDLINE inquiries by mail.[22]

Shortly before an annual international technical symposium sponsored by the Society of Photo-Optical Instrumentation Engineers (SPIE) occurred in San Francisco during mid-August 1982, the Department of Defense intervened to prevent the scheduled presentation of approximately 100 professional papers. The withdrawn items—about 700 papers were planned for the conference—reportedly covered such advancements in optical technology as small mirrors for laser communications between satellites and submerged submarines, infrared optics, airborne reconnaissance sensors, and other microelectronic research. Pentagon officials argued that the disputed papers were subject to control under the International Traffic in Arms Regulations and DOD contract review authority. A sizable number of the presentations apparently derived from research sponsored by the Department and, although cleared earlier by DOD program managers, were subsequently restricted by higher officials within the Pentagon. It was uncertain how many of the withdrawn papers actually contained information sufficiently sensitive to warrant their suppression, but there were a number of foreign nationals scheduled to attend the gathering, which reportedly was one of the reasons for DOD intervention.[23]

In testimony before a subcommittee of the House Committee on the Judiciary, F. Karl Willenbrock, a professor of engineering at Southern Methodist University and the chairman of the Technology Transfer Committee of the Institute of Electrical and Electronics Engineers (IEEE), related three other incidents that occurred shortly after the SPIE conference controversy. Discussing the first two of these situations, he began with an IEEE-sponsored international gathering in Philadelphia in November 1982:

> Five days before the conference was to open and after the conference publication had been printed, an official of Texas Instruments, Inc. requested, by telephone, that three papers written by TI engineers on Very Large Scale Integrated (VLSI) circuits be withdrawn because the U.S. Air Force Systems Command considered the release of these papers to be potentially damaging to U.S. interests. The authors, who are all IEEE members, had believed their papers were cleared having followed the usual procedures. Apparently the government reviewers did not decide until very late, that is until five days before the conference and after the conference digest had been printed and copies supplied to reporters, that the papers should not be published. The conference managers were asked to excise the papers from the already printed conference digest and to ask reporters who had received prepublication copies to return them. The conference managers required that before taking these steps a written explanation of the reasons for removing the papers be given; that the authors themselves request removal; and that an agreement be made to

> pay the costs of destroying portions of the already printed record. In response to these requirements, a review of the papers by the Air Force resulted in a decision that the original papers could be presented as planned. Another incident occurred in [September] 1982 in connection with the EASCON [Electrical Aero Space Conference] conference held in Washington and sponsored by the IEEE Aerospace and Electrical Systems Society. Just before the conference opened, the chairman was asked by an Air Force representative to destroy all conference records and to cancel the presentation of certain papers. The conference chairman responded that he might agree to do so if the estimated cost of between $25,000 to $50,000 were borne by the Air Force. A day later, the Air Force representative withdrew his request. Later, a Navy representative made a similar request which also was withdrawn after the costs were described.[24]

The third case described by Dr. Willenbrock involved a manuscript that was submitted by an external author for publication in IEEE's magazine, *Spectrum*, sometime in October 1982.

> The article, entitled "Out-Numbered and Out-Weaponed by Soviets, the U.S. Army Shoots for High Technology," was subjected to the usual expert review process. Since the author quoted the Secretary of the Army, the article was sent to his office for review. A month and a half later, the *Spectrum* staff editors received a telephone call from the Army Office of the Chief of Public Affairs with the message that the manuscript contained classified information and should be shredded immediately. Upon questioning, the Army representative identified three statements as being classified. The *Spectrum* staff investigated the origins of these three statements and found that two statements had been published in an unclassified widely disseminated Army publication entitled "1982 Weapons Systems," and the third was from public testimony of the Army Chief of Staff to the 97th Congress. When this information was presented to the Army representative, it was agreed that the two published phrases were not really classified, but that the Chief of Staff's testimony had been reclassified. The explanation was offered that sometimes unclassified information can be put together in such a way as to be reclassifiable. The issue was not pursued further since the article did not meet the technical standards appropriate for *Spectrum*.[25]

In July 1983, prior to the convening of the Fourth International Conference on Permafrost in Fairbanks, Alaska, Stephen D. Bryen, Deputy Assistant Secretary of Defense for International Trade and Security Policy, ordered the withdrawal of six papers that had been submitted for presentation at the meeting and publication in the proceedings. The suppressed papers were based on research completed at the United States Cold Regions and Engineering Laboratory of the

Army Corps of Engineers. Although it was reported that the six presentations involved no classified information, and no official reason for their withdrawal was offered at the time, Bryen subsequently explained that there were "several Russians at the conference" and that the particular papers in question could have provided useful information concerning aspects of maintaining certain military facilities in cold climates, such as Siberia.[26]

Later, during the annual meeting of the American Vacuum Society in early November, FBI agents arrested an East German physicist, Alfred Zehe, on espionage charges. Although the Boston arrest was the culmination of a two-year investigation and Bureau officials acknowledged that Zehe had spent very little time at the AVS meeting, the incident was unsettling for Society leaders and scientists. There was some concern that the Reagan Administration would use the case to justify additional controls on open scientific professional conferences and symposia attended by foreign nationals. Furthermore, there was some anger over the methods used by the FBI in making the arrest. AVS president John Arthur, an Oregon State University physicist, said that a man, identifying himself as a Bureau agent, had appeared at Society offices about a week before the annual meeting was scheduled to occur and had requested complimentary entry badges in order that six agents might attend the meeting under false names. AVS office staff complied with this request, but did not inform Arthur or other members of the executive committee of their action. When Arthur learned of this development just before the conference began, he contacted the FBI and indicated that the gathering was open to anyone paying the registration fee. Later, a man presenting FBI credentials paid to register himself and five other individuals. They subsequently attended the AVS meeting and arrested Zehe at the conference hotel in the afternoon. Several days later, the FBI requested a list identifying the 2,600 attendees at the Society meeting. AVS officials refused to comply unless presented with a subpoena. Furthermore, some Society officers were convinced that, in conjunction with the Bureau's investigation, their telephones had been tapped.[27]

The presidents of Stanford, CalTech, and MIT jointly informed the White House in early April 1984 that their universities would refuse to conduct certain kinds of sensitive, but unclassified, scientific research for the Pentagon if DOD reviewers were given the power to restrict the publication of findings. The warning was in reaction to proposed Defense Department policy, to be implemented through contract provisions, authorizing research managers to prohibit the publication of some kinds of study findings and comment in advance on the propriety of publishing results in other categories.[28] These were not, however, the only university leaders troubled by the proposed new DOD policy. Soon, their cause received influential support from within the Pentagon. Richard DeLauer, Under Secretary of Defense for Research and Engineering, not only distanced himself from the proposed DOD policy, but also took issue with the basic idea of a special category of unclassified research that required restriction.[29]

Indications were given in May that DeLauer's position was gaining acceptance within the Pentagon and the administration. In testimony before subcommittees of the House Committee on Science and Technology, DeLauer's deputy, Edith Martin, revealed a draft policy statement making security classification the primary means of controlling the publication of DOD-sponsored "fundamental research." Asked what constituted "fundamental research," which was undefined in the draft statement, Martin described it as "a scientific endeavor that expands our basic knowledge and understanding of the world and the universe. In applied research," she commented, "we have taken that understanding and given it some specific use or meaning."[30]

Later, in an October memorandum to Pentagon research chiefs, DeLauer prescribed that, "consistent with existing statutes, no controls other than classification may be imposed on fundamental research and its results when performed under a federally supported contract." Furthermore, the memorandum specified that, for DOD, "the decision whether a particular research activity is or is not fundamental will be determined primarily by considering the following easily identified characteristics: (1) performer (for example, university, industry, in-house), (2) budget category (for example, 6.1, 6.2 [pertaining to basic research]), (3) sponsoring DoD entity, (4) special contractual provisions."[31] Many in the university research community were pleased with this policy and the very similar policy expressed in NSDD 189, which was the formal version of Martin's draft policy statement. Others, however, realized that these policy statements, with their references to existing or applicable statutes, such as export laws, probably did not actually provide much relief from national security controls.

In June 1984, it was revealed that the Navy Material Command did not want its personnel actively participating in other than DOD-sponsored symposia, conferences, or other similar forums, including campus teaching, on weapons and associated technologies. Four Navy civilian employees were particularly singled out by name for unfavorable mention. All were instructors in radar courses given at the George Washington University continuing education program. Their teaching activity was seen as "another example of the type of thing that contributes to the undesirable availability of sensitive information on Navy programs and capabilities." The director of GWU's continuing engineering education program responded, saying that the university "would mutiny if we did teach anything classified," that instructors were advised to be careful about revealing sensitive information in the classroom, and that a ban on such teaching by military personnel would ultimately be more detrimental to the Pentagon than to the universities.[32]

ENTER SELF-CENSORSHIP

Another phenomenon concerning scientific communication also became noticeable during 1984—self-censorship. For example, a composite materials symposium sponsored by the National Aeronautics and Space Administration and DOD in conjunction with the American Ceramics Society's Eighth Annual Conference on Composite and Advanced Materials, which was held in January, was restricted to "U.S. citizens only."[33]

In May 1984, individuals attending the Twenty-Fifth Structures, Structural Dynamics and Materials Conference found that two of its sessions were closed to foreigners. The conference program carried a notice citing export regulations as the basis for the restriction, and advising those seeking admission that "you must have with you a document signed by your company security officer (e.g.: company badge which specifically states U.S. citizenship), birth certificate, naturalization papers, passport, or voter registration card in order to certify that you are a U.S. citizen. If you fail to carry this citizenship material with you," it warned, "you will be denied admission to these sessions."

This limitation occurred even though one of the conference sponsors, the American Institute of Aeronautics and Astronautics, had a policy of holding open, public meetings if at all possible. Reporting on the incident, the privately published newsletter *Defense Week* commented: "Never before have organizers volunteered to close a meeting unless it was jointly sponsored by the federal government."[34]

The Society for the Advancement of Material and Process Engineering sponsored an October conference, *High-Tech Review—1984*, where metal matrix and carbon-carbon discussion sessions were available only to those providing proof of U.S. citizenship.

Furthermore, both the University of Maryland at College Park and the University of California at Los Angeles were willing to restrict a short course on metal matrix composites. Taught in cooperation with the Metal Matrix Composites Information Analysis Center—a DOD affiliate—the class was open only to American nationals because some composites are export-controlled.[35]

Some scientific groups and universities, having observed or experienced the more vigorous application of national security controls to traditional scientific communication, probably entered into self-censorship to avoid disputes and conflicts with federal agencies enforcing the restrictions.[36] Furthermore, the practice seemingly gained acceptance within some sectors of the scientific community.[37] Robert Park, executive director of public affairs for the American Physical Society and a University of Maryland physicist, noted in the fall of 1986 that:

> A number of societies continue to restrict sessions and in some cases entire conferences to United States citizens only, in spite of assurances from officials of the Defense Department that the action is unnecessary. The impetus for the restrictions often comes from potential authors of contributions. A residue of doubt in the mind of the researcher or his or her management as to whether results may be subject to government controls is often sufficient to inhibit its release in an open forum. Thus, self-censorship resulting from past intimidation serves to extend the effect of controls.[38]

CONTINUED CONFLICT

Disputes over national security controls on traditional professional communication by campus-based scientists continued as the Reagan Administration entered its second term. A major controversy arose in April 1985, when DOD identified 43 papers submitted for presentation at a Washington meeting of the Society of Photo-Optical Instrumentation Engineers as being unsuitable for open sessions. Once again, the criteria and authority for restricting scientific papers derived largely from the national security provisions of the Export Administration Act, but the studies of some contributors arose from DOD-sponsored research. Subsequently, over half of the disputed papers were satisfactorily modified to allow their presentation. Some of the remaining papers were withdrawn by their authors, and efforts to have others given at special closed sessions became mired in bureaucratic problems. It would appear that the incident generated especial consternation due to the quantity of research studies involved, the pursuit of closed sessions as a readily acceptable remedy, and reliance on authority that, many thought, provided protection against automatic disclosure under the Freedom of Information Act, but no regulatory power.[39]

During the summer of 1985, officials from the Department of Defense and the Department of State were pressuring the National Science Foundation (NSF) and the four campus supercomputer centers it was establishing to set restrictions on access to these facilities by foreign nationals. At the time, NSF had already completed supercomputer contracts with Princeton University and the University of California at San Diego that contained no such limitation provisions, but did commit them to whatever federal policy was later adopted on the access question. The University of Illinois and Cornell University had not yet signed their contracts. At a minimum, NSF was uncomfortable with the suggested restriction and the universities were concerned about the threat this caveat posed for academic freedom.[40] Illinois and Cornell later signed their supercomputer contracts, which were devoid of access restraints. The matter of restrictions, however, was not settled; the prospect of subsequent supercomputer access limitations left open the possibility that contracts, such as those of Illinois and Cornell, might have to be renegotiated.[41]

In the summer of 1985, the Departments of Commerce and Defense also evidenced renewed interest in possibly subjecting biotechnology to export controls. Earlier, in the spring of 1984, a Pentagon intelligence official had warned scientists that without export regulation, American advances in biotechnology might be "turned against us" by the Soviet Union in exotic new germ warfare weapons.[42] About a year later, it was revealed that DOD would ask the Coordinating Committee for Multilateral Export Controls (COCOM)—an international export control coordinating body largely consisting of the U.S. and its NATO allies—to place exports related to genetic engineering on its watch list and begin monitoring these transactions. Academic and industry scientists were immediately concerned that this request was a prelude to U.S. national security export restriction of the flow of biotechnology research information. Officials from the Pentagon and the Office of Science and Technology Policy subsequently gave assurances that no restraint of scientific communication was being discussed. Nonetheless, biotechnology remains an area of considerable interest in terms of export control.[43]

Controversy erupted again, in August, over a memorandum to top DOD research officials from James Ionson, an innovative science and technology (IST) director in the Pentagon's Strategic Defense Initiative Office. The purpose of the memorandum was to indicate that no restrictions would be placed on the publication of campus-based research sponsored by Ionson's entity, unless agreed to in advance by contract. Indeed, the memorandum stated such a policy position, but also indicated something more. The crucial paragraph said:

> Decisions regarding publication of the results of unclassified IST research performed on university campuses will normally be the responsibility of the university author. However, where there is likelihood of disclosing

> operational capabilities and performance characteristics of planned or developing military systems, or technologies unique and critical to defense programs, the contract will stipulate that the responsibility for the release of information resulting from IST research belongs to the sponsoring office.[44]

In September 1985, the presidents of a dozen scientific professional societies sent a letter to Secretary of Defense Caspar Weinberger "to bring to your attention our shared concerns regarding new controls being applied by the Department of Defense to unclassified technical information." Such controls, in the view of the signatories, "have resulted in disruption of technical conferences and conference planning and have forced the introduction of a new category of technical session, one that is unclassified but with limited access." The presidents indicated that it was the responsibility of the government, not their societies, to implement these controls. "Therefore," they said, "our organizations *will not be responsible for, nor will they sponsor, closed or restricted access technical sessions at meetings or conferences conducted under their auspices.*"[45]

The effect of this letter is difficult to assess. It expressed an unwillingness by some major professional societies within the American scientific community to engage in self-censorship or to police their meetings for possible breaches of national security law. Its transmittal also coincided with an end of the disruption of professional conferences by federal agencies to restrict presentations or participants for reasons of national security.[46] There was, however, no cessation of government attention to the acquisition of sensitive, but unclassified scientific and technological information by foreign interests to the detriment of the national security of the United States.

During the latter years of the 1980s, for example, there was governmental concern that sensitive, but unclassified scientific and technological information in electronic formats might be particularly vulnerable. As early as the spring of 1984, a newspaper account revealed that the Reagan Administration was considering expanded electronic surveillance activity to prevent the smuggling of computer software and data, whether campus or commercially generated, through international telephone calls.[47] During the summer of 1986, an Air Force proposal to prevent direct access by noncitizens to unclassified electronic databases was generating strong opposition. As envisioned, the restriction would have applied to both government and private databases, including commercial services such as Lockheed Corporation's Dialog, which links with systems maintained by scientific societies and universities.[48]

Attempts to place national security controls on subscribers to commercial databases continued during the final years of the Reagan Administration. A particularly ambitious effort was made in 1986, through NSDD 145, to give the National Security Agency authority for prescribing computer and telecommunications security standards for federal agencies, contractors, and private sector enterprises. Many in Congress regarded this a dangerous and unwarranted policy action that encroached upon the legislature's constitutionally prescribed lawmaking authority, assigned regulatory power to a quasi-military agency that more properly should be exercised by a civilian entity, and otherwise assumed a need to control "sensitive, but unclassified" information held in private sector databanks. Consequently, NSDD 145 was largely displaced by the Computer Security Act of 1987, which gave the National Bureau of Standards, now the National Institute of Standards and Technology, responsibility "for developing standards and guidelines for Federal computer systems, including responsibility for developing standards and guidelines needed to assure the cost-effective security and privacy of sensitive information in Federal computer systems."[49] This issue is discussed further in the next chapter.[50]

Stepped up investigative activity at about this same time brought a Federal Bureau of Investigation (FBI) counterintelligence program to public attention. Inaugurated some 20 years earlier, the Library Awareness Program gained visibility and notoriety in September 1987 when the *New York Times* disclosed that FBI agents were asking city librarians to watch for and report on users who might be seeking technical information allegedly harmful to national security. Persons of a foreign character were of particular interest to the Bureau. Librarians also were to be on the watch for spy-recruiting activities by such individuals.[51] Alarmed by the Bureau's vague standards for identifying suspects and the threats posed to both intellectual freedom and personal privacy—the confidentiality of library records is protected by law in 39 states—library professional associations slowly discovered that the New York City effort was not an isolated case.[52] In June and July 1988, the House Subcommittee on Civil and Constitutional Rights, which has long monitored FBI operations, held oversight hearings on the Bureau's counterintelligence visits to libraries. These proceedings gave the library community an opportunity to express outrage over the Bureau's intrusions, disregard for pertinent State confidentiality laws, and clumsy attempts to restrict open collections; they also provided an opportunity for congressional overseers to indicate displeasure with much of this FBI activity, which subsequently subsided.[53]

CAMPUS RESPONSES

Encroachments in the name of national security upon the research and professional communications of campus-based scientists did not occur without protest, rejoinder, and attempts at reconciliation. As noted above, in early 1981, the presidents of Cornell, MIT, CalTech, the University of California, and Stanford, jointly expressed their displeasure to the Secretaries of Commerce, State, and Defense concerning efforts to apply federal export regulations to universities. Similarly, that fall, various academic leaders declined to comply with a State Department request that visiting students from the People's Republic of China be monitored and otherwise restricted from engaging in research and studies in certain scientific and technological areas that were regarded as being subject to export controls.

Some professional organizations—notably the American Association for the Advancement of Science, the Association of American Universities, the American Physical Society, and the Institute of Electrical and Electronics Engineers—also indicated their disapproval of increased national security controls on traditional scientific communication, but sought, as well, to promote dialogue and understanding about the values and interests at issue in the controversy. AAAS, for example, through a resolution adopted by its governing council, went on record as opposing "governmental restriction on the dissemination, exchange, or availability of unclassified knowledge." This position was taken at the time of the Association's 1982 annual meeting where a panel, including Deputy Director of Central Intelligence Bobby R. Inman, discussed and debated new national security controls, as well as other aspects of scientific freedom and national security.[54] Other discussion forums, information exchanges, and studies concerning national security controls and secrecy were subsequently sponsored by the Association. Indeed, during the decade, the AAAS Committee on Scientific Freedom and Responsibility devoted a considerable amount of its resources to monitoring, assessing, and publicizing the impact of increased national security controls on traditional scientific communication.

A January 1982 Defense Science Board (DSB) report elicited a different kind of response from the universities.[55] Prepared by the DSB Task Force on University Responsiveness to National Security Requirements, chaired by Dr. Ivan Bennett, Jr., executive vice president for health affairs at New York University and provost and dean of the New York Medical Center, the report was to address university interest in and ability to support national security requirements in both manpower training and basic research, the adequacy of DOD research contracting procedures and role relative to the support of basic research by other agencies, the impact of export controls on the conduct of research and teaching activities of universities, and the problems caused by the high percentage of foreign nationals in science and engineering. The published evaluation clearly recognized that the Pentagon was in a precarious position when trying to control communication of sensitive, but unclassified research. "DoD is caught in a dilemma," it said. "If it vigorously attempts to regulate the flow of scientific information in the scientific community, it could jeopardize the strength and vitality of the very community it is seeking to revitalize for the sake of national defense. On the other hand," it was noted, "if DoD abandons any attempt at regulation in the university context, it could seriously compromise and, in certain cases, totally undercut other efforts to control the out-flow of militarily critical technology."[56]

The Task Force acknowledged that a middle ground position "is a difficult one to establish," but, nonetheless, it "attempted, if not to solve the problem, to at least lay a framework for solving the issue by means both practicable and, it is hoped, acceptable to the academic community."[57]

Indeed, the report did not solve "the problem" due, in part, to the manner in which the Task Force perceived and formulated it. For example, saying "university research that DoD would consider militarily critical [and, therefore, subject to control] is for the most part DoD-funded," was irritating because of the implication that national security controls could be applied to research in which the government had no proprietary interest.[58] Elsewhere, the report, without offering any substantiation, made the following inflammatory allegation: "Numerous advances in Soviet military weapon systems are directly traceable to technology transfers that occur as a result of Soviet and Warsaw Pact student and scientist exchanges and their attendance at international scientific symposia held in the United States."[59]

Among its findings, the Task Force held that "(c)ertain specific areas of university research, especially those conducted under DoD contract, are sensitive from an export control point-of-view," and, moreover, "DoD cannot exempt the universities from the legislated restrictions of International Traffic in Arms Regulations (ITAR) and the Export Administration Regulations (EAR)." Interpretation of these regulations for university researchers, said the report, "is required."[60]

However, although statements such as these did not solve "the problem," the Task Force did offer a recommendation that proved acceptable to the academic community. The report proposed that the Under Secretary of Defense for Research and Engineering "create a forum to allow periodic consultations between senior university representatives and DoD officials on the full range of research-related needs and issues that affect the Department's ties with universities."[61]

Shortly thereafter, in February 1982, the DOD-University Forum was launched, largely at the instigation of Under Secretary of Defense Richard DeLauer and Stanford president Donald Kennedy, who served together as the co-chairs of the panel. An informal body composed of eight university presidents, the heads of three educational organizations, and nine representatives from Defense agencies, the Forum's initial deliberations were devoted to technology transfer and export control considerations, research support for universities, and graduate education in the physical sciences. After almost two years of frequent meetings, the ad hoc, quasi-official Forum was chartered as an official DOD advisory committee on December 15, 1983.[62]

In the summer of 1987, the Association of American Universities (AAU), one of the educational organizations formally represented on the Forum, released a collection of publications that were prepared by the AAU for the Forum. Sets of these materials were distributed to approximately 200 universities and scientific societies. Each set included a compilation of statutes, regulations, and policy statements bearing upon national security control of scientific communication, a related explanatory brochure and poster, and a book of background readings. No similar publications were subsequently offered by the Forum and it appears that the advisory panel's role diminished in succeeding years.

Another attempt to reconcile differences between campus-based scientists and federal officials exerting more stringent national security control over traditional scientific communication was made by the National Academy of Sciences. Through the Academy's Committee on Science, Engineering, and Public Policy, a special study panel was created in late March 1982, to examine "the relation between scientific communication and national security in light of the growing concern that foreign nations are gaining military advantage from such research."[63] Chaired by Dale R. Corson, a physicist and president emeritus of Cornell University, the Panel on Scientific Communication and National Security was composed of 19 members, including senior university faculty members and administrators, former federal agency officials, and leaders in high-technology industrial firms.[64] Among the sponsors funding the study were the Department of Defense, the National Science Foundation, the American Association for the Advancement of Science, the American Chemical Society, the American Geophysical Union, and the National Academy of Sciences.

Through arrangements jointly made by panel chairman Dale Corson, Academy president Frank Press, and DOD Under Secretary Richard DeLauer, all members of the study panel not having such status were granted security clearance to enable receipt of "classified information about technology transfers to other countries." Furthermore, the panel "was subsequently given three secret-level briefings by members of the intelligence community. In addition," according to the final report, "a subpanel, comprising six members of the Panel who [held] clearance at the highest level, was briefed at two additional meetings." In becoming privy to such classified information, the Corson panel, nonetheless, did not want to use it in any way that would cause the study group's final report to be subjected to official secrecy. However, "a classified version of the subpanel report based on the secret intelligence information it was given" did result, and this document was made "available at the Academy to those with the appropriate security clearance."[65]

Among the other information resources utilized by the study panel were briefings from representatives of the Departments of Commerce, Defense, and State, and from representatives of the intelligence community, including the CIA, the FBI, the Defense Intelligence Agency, and the National Security Agency. Various expert views were solicited and special studies were commissioned. Finally, during its deliberations, discussions, and briefings, the Corson panel had the invited participation of liaison representatives from the study's sponsoring organizations, as well as from the Departments of Commerce and State, the Office of Science and Technology Policy, the federal intelligence community, the Association of American Universities, the Institute of Electrical and Electronics Engineers, and the American Physical Society.[66]

The study panel released its final report, which had the support of all of its members, at the end of September 1982.[67] However, it was not greeted with universal approval by the scientific community.[68] The report generally conceded "a substantial transfer of U.S. technology—much of it directly relevant to military systems—to the Soviet Union from diverse sources," but concluded "that universities and open scientific communication have been the source of very little of this technology transfer problem."[69] The report, noting that "current proponents of stricter controls advocate a strategy of security through secrecy," also offered an alternative view of the relationship between science and national security. "In the view of the Panel," said the report, "*security by accomplishment* may have more to offer as a general national strategy. The long-term security of the United States," it continued, "depends in large part on its economic, technical, scientific, and intellectual vitality, which in turn depends on the vigorous research and development effort that openness helps to nurture."[70] The report particularly recognized the importance of traditional openness in the basic research of campus-based scientific enterprise.

> Basic research investigations undertaken today may lead to applications in the long term (perhaps 10 to 20 years from now), often in unexpected ways. To attempt to restrict access to basic research would require casting a net of controls over wide areas of science that could be extremely

> damaging to overall scientific and economic advance as well as to military progress. The limited and uncertain benefits of such controls are, except in isolated areas, outweighed by the importance of scientific progress, which open communication accelerates, to the overall welfare of the nation.[71]

The Corson report prefaced its more specific recommendations with at least two general suggestions. The first was directed at the entire current control system.

> At this stage, the government's technology transfer controls have a very wide compass. There is a risk that by covering such broad expanses of technology and by implicating the scientific activities across the same range, the overall efficiency of the effort will suffer. Two principles can be applied to bring a more coherent focus to the problem. First, the government should concentrate on the most *feasible forms of control* and should eschew regulations that impose compliance burdens without significantly affecting leakage. Second, it should concentrate its resources more systematically on those *technologies that are of greatest relevance* to near-term Soviet military strength.[72]

This was the panel's celebrated entreaty for "tall fences" around a limited number of narrow areas of research. Its other general suggestion was offered in the form of a guiding conclusion: "(1) that most university research should be unrestricted; (2) that in rare cases such research should be classified for reasons of national security; and (3) that in a few specific cases, limited control measures short of outright classification may be warranted."[73]

Among the more specific recommendations of the Corson report were proposals that:

- "no restrictions of any kind limiting access or communication should be applied to any area of university research, be it basic or applied, unless it involves a technology meeting" all four specified criteria concerning its sensitivity;
- "if government-supported research demonstrably will lead to military products in a short time, [security] classification should be considered" to protect such research;
- "in the limited number of instances in which all of [its] four criteria are met but in which classification is unwarranted, the values of open science can be preserved and the needs of government can be met by written agreements" conforming with certain specified principles, including abstinence from applying export controls;
- "unclassified information that is available domestically should

receive a general license (exemption) from the formal [export] licensing process";

- "information that is not directly and significantly connected with technology critical to national security should also receive a general license (exemption) from the formal [export] licensing process";
- there be "a drastic streamlining of the [Militarily Critical Technologies List] by reducing its overall size to concentrate on technologies that are truly critical to national security"; and
- various actions be taken by government officials, university administrators, and faculty to assure that academic and scholarly exchanges are not misused by foreign nationals for intelligence or improper technology transfer purposes.[74]

PURSUING CORSON

Chairman Corson had met earlier in the week with Secretary of Defense Caspar Weinberger to discuss the panel's findings and recommendations, and the President's national security adviser, William Clark, attended an Academy briefing held prior to the public release of the final report.[75] In his remarks at the Academy's press conference on the study, Corson indicated that Clark had spoken "supportively" about the panel's report, and would brief the President on it. Furthermore, he had promised to have the National Security Council staff, in Corson's words, "work on examining the report and following it up."[76] It should be remembered, however, that Clark purported "not to be a policymaker but to be a lawyer insuring that all arguments were put before his one-man court, the president." Furthermore, although this posture was seen as something of a sham, one of the things that Clark actually did—significantly, with regard to national security controls—was encourage President Reagan's anticommunist hard line.[77]

During its deliberations, the Corson panel was not given the opportunity to meet with two powerful proponents of increased national security controls on scientific communication. Richard N. Perle, Assistant Secretary of Defense for International Security Policy, referred the panel to his deputy, Stephen D. Bryen, DOD Deputy Assistant Secretary for International Economic, Trade, and Security Policy. He reportedly met with panelists once in April, became angry during the session, and refused to further assist them. At the Commerce Department, Lawrence J. Brady, Assistant Secretary for Trade Administration, had Bohdan Denysyk, Deputy Assistant Secretary for Export Administration, stand in for him when the Corson panel sought an audience.[78]

As October disappeared and November started to fade, concern began mounting that the Reagan Administration, regardless of the initiatives taken by Clark, might ignore the Corson report. Perhaps Perle and Brady had stifled White House interest in it. Then, in December, prospects brightened. Two days before Christmas, President Reagan signed National Security Study Directive 14–82 creating an interagency group to "review the issue of protecting sensitive, but unclassified scientific research information, taking into account the recommendations made by the National Academy of Sciences' Panel on Scientific Communication and National Security (the Corson Panel)."[79] The NSSD established the terms of reference for the review and its objective—production of a national security decision directive setting policy for the protection of sensitive, but unclassified, scientific information. Responsibility for chairing the interagency group, as well as scheduling and managing its study, was given to the Office of Science and Technology Policy (OSTP). For some within the scientific community, hopeful of a bit of relief from the new stringent national security controls on their professional communication, such OSTP leadership seemed promising. This was an agency that should be sympathetic to scientists' tradition of and needs for openness. Furthermore, at the time of its issuance, the Corson report, according to one press account, had been "greeted warmly" by George A. Keyworth II, OSTP's director and President Reagan's science adviser. "The last thing we want to do," he had commented, "is ape the repressive Soviet model, which stifles technological innovation through its obsession with secrecy."[80]

Others, however, were less assured by OSTP's leadership role. Any NSDD drafted by the interagency group would be subject to refinement and final approval within the National Security Council. Keyworth and his staff had little entree there, but for defense officials like Perle, it was very accessible and familiar ground.

NSSD 14–82 tasked the interagency group to report to the National Security Council no later than March 1, 1983. This deadline was missed and the entire review effort eventually collapsed, largely due to three successive turnovers in the interagency group's leadership. A high degree of secrecy surrounding the review effort also disenchanted various quarters of the scientific community. In February, the White House, pursuing a broad study of export controls, began considering the development of its own draft NSDD on the transfer of scientific, technical, and engineering information. Perhaps it was unhappiness with OSTP's management of the Corson report review that prompted the White House to bypass the agency when preparing the administration's legislation revising and extending the Export Administration Act.[81]

In November 1983, approximately a year after the issuance of the Corson report, the NAS Committee on Science, Engineering, and Public Policy (COSEPUP), which had been instrumental in establishing the Corson panel, called for an update on the Reagan Administration's response to the report and related events. This evaluation was subsequently prepared by Lawrence E. McCray and Mitchel B. Wallerstein, who held positions within the NAS complex and had been the Corson panel's principal staff assistants. It was presented to COSEPUP in late January, and represented only the views of the two authors.[82] A version of it later appeared in *Science* magazine.[83] Both accounts concluded that

- the "government has not found it possible to act in a manner compatible with the major Corson Report principles";
- the "continuing lack of effective government-wide coordination raises important risks, including disparate agency policies that do not adequately balance national goals, wasteful allocation of national resources among programs of varying effectiveness," and "confusion and skepticism in the research community"; and
- "there is little progress toward an improved objective understanding of the technology leakage problem and the effects of control measures."[84]

The original COSEPUP evaluation, however, also offered two additional conclusions.

> (1) For most aspects of the technology transfer problem, we seem to have moved from a period of public controversy to a more subdued period of policy development and implementation. However, continuing incidents regarding international scientific conferences remain a source of current acrimony.
>
> (2) There is little doubt that one useful development since the Corson Panel was established has been an enhanced ability to distinguish between the general problem of unwanted technology transfer from all sources and those aspects that pertain to the open communication of scientific information. Despite the possibility of increased targeting [by hostile intelligence forces] in the future, unclassified scientific communication continues at present to be a small part of the overall problem.

Nonetheless, White House efforts at perfecting a NSDD on sensitive, but unclassified scientific and technological information were progressing. A draft was revealed by Edith Martin, Deputy Under Secretary of Defense for Research and Engineering, at a late May hearing jointly held by two subcommittees of the House Committee on Science and Technology.[85] Martin, who was the deputy to Under Secretary DeLauer, indicated that the draft directive had been developed with DeLauer's support, with assistance from the DOD-University Forum, and in response to the Corson panel's recommendation concerning the safeguarding of gray areas of research.[86] Although subcommittee members largely gave the draft a positive reception, there was some concern that it did not define the "fundamental research" it was addressing.[87] In October, a DeLauer memorandum to the principal research officials of the Pentagon not only provided such a definition, but also indicated that the policy expressed in the draft was "to be applied consistently to all DoD sponsored research." This policy, as stated in the draft, was "that the mechanism for control of fundamental research in science and engineering at colleges, universities and laboratories under contract to U.S. Government Agencies is [security] classification." However, it also added that, "Consistency of this policy with applicable U.S. Statutes must be maintained."

Although the draft remained under consideration by the National Security Council for almost a year, it finally emerged in September 1985 as NSDD 189.[88] Its policy statement was worded somewhat differently than the draft, but contained the same primary premise *and* the same problematic caveat:

> It is the policy of this Administration that, to the maximum extent possible, the products of fundamental research remain unrestricted. It is also the policy of this Administration that, where the national security requires control, the mechanism for control of information generated during federally-funded fundamental research in science, technology and engineering at colleges, universities and laboratories is [security] classification. Each federal government agency is responsible for: a) determining whether classification is appropriate prior to the award of a research grant, contract, or cooperative agreement and, if so, controlling the research results through standard classification procedures; b) periodically reviewing all research grants, contracts, or cooperative agreements for potential classification. No restrictions may be placed upon the conduct or reporting of federally-funded fundamental research that has not received national security classification, except as provided in applicable U.S. Statutes.

Although, NSDD 189 stressed that security classification would be the only basis for restricting federally-funded fundamental research information, it made an exception for protective authority provided in statutory law. In brief, it set no limitation on continued resort to the Export Administration Act, the technical data safeguard authority of the Secretary of Defense, or the Invention Secrecy Act. Moreover, this policy seemed to have little or no relation to the recommendations of the Corson panel.

CONGRESSIONAL CONSIDERATION

When the Corson panel's final report was issued in September 1982, copies were widely distributed on Capitol Hill, but there was no immediate formal response from Congress. Chairman Corson did not appear before a congressional panel to discuss the report until May 1984, when the NSDD 189 draft was unveiled at the same proceeding.[89]

Congress, however, at the outset of the decade, had been made aware of problems campus-based scientists were having with national security controls on their research and professional communication. During February, March, and August 1980, a subcommittee of the House Committee on Government Operations had explored difficulties arising from invention secrecy, cryptography research restraints, and atomic energy data restrictions.[90] Subsequently, campus-based scientists would have their opposition to increased national security controls aired before subcommittees of the House Committee on Science and Technology[91] and the House Committee on the Judiciary.[92] Such views, as noted previously, were not formally presented at hearings by congressional panels reauthorizing the Export Administration Act during the 98th, 100th, or 101st Congresses. In view of the manner in which the House and the Senate rely on their committees to structure and organize their legislative work, these reauthorizing panels were crucial forums for pursuing relief from at least one kind of national security control.

The Export Administration Act was reauthorized in July 1985.[93] NSDD 189 was issued in September. Self-censorship at scientific professional conferences and symposia had been evident since the beginning of the year. This combination of developments made it difficult to assess the extent to which increased national security controls on professional communication continued to be problematic for campus scientists.

Although there appeared to be a subsiding of public criticism and protest of the national security restraints, there were some important indications that the general problem had not been resolved. Serious studies of the U.S. and multilateral national security export controls continued to appear, with recommendations and strategies to regulate international technology transfer. These assessments—produced by the Center for Strategic and International Studies in 1985,[94] the Business-Higher Education Forum in 1986,[95] a National Academy COSEPUP panel in 1987,[96] the Computer Science and Technology Board of the National Research Council in 1988,[97] and another Academy panel in 1991[98] —moved beyond the controls on campus-based science to technology transferred as part of or in association with commercial activities.[99] Furthermore, they reflected a growing concern that national security export controls were unnecessarily restraining American business competitiveness in international markets.

Congress responded to this concern in 1988 with the Omnibus Trade and Competitiveness Act, which, as the title suggests, sought to improve the U.S. competitive position in a number of policy adjustments, including some relaxation of national security export restrictions.[100] Exports were promoted by a number of general enhancements and by amendments to the Export Administration Act eliminating licenses for shipments to certain countries of goods that could currently be exported to the People's Republic of China without special approval, for exports to certain countries judged to have effective export control systems, and for shipments to non-Soviet bloc countries of low-technology goods that could currently be exported to Communist bloc nations without special approval. Re-export licenses for items sold to some U.S. allies were eliminated as well. Also struck from the list of controlled goods was technology that could be exported to Soviet bloc countries without special approval, except for items on which the U.S. and certain of its allies agreed to maintain notification requirements. In addition, the Secretary of Commerce and the Secretary of Defense were obligated to "enter into appropriate arrangements with the National Academy of Sciences and the National Academy of Engineering . . . to conduct a comprehensive study of the adequacy of the current export administration system in safeguarding United States national security while maintaining United States international competitiveness and Western technological preeminence."[101] This study did not get underway until late the following year and, therefore, was not ready, as was the original intention, for the reauthorization of the Export Administration Act in 1990.

During the final year of the Reagan Administration, congressional efforts at promoting competitiveness while reducing national security export restrictions were facilitated if not aided by the absence of hardline export control stalwarts, such as Weinberger and Perle, and by dramatic political developments in the Soviet Union that contributed to a significant reduction of U.S.-U.S.S.R. tensions.

However, the new President, George Bush, exhibited a very cautious approach to the changing Soviet circumstances during most of 1989, his first year in office. Defense and foreign policy responses to Soviet President Mikhail Gorbachev's proposals were guarded, and the United States remained largely in a reactive posture, making almost no reciprocal initiatives. The new Secretary of Defense, Dick Cheney, began articulating views in late April indicating little confidence in the durability of Gorbachev or his reforms and concern that the Soviet Union would consequently revert to its prior threatening stance—a position he continued to hold.[102] Vice President Dan Quayle evidenced similar attitudes, warning, for example, in October 1989, that the Soviet Union "remains our potential adversary,"[103] and two months later, after the Malta summit, that Soviet foreign policy had not changed. The United States was "still dealing with a totalitarian government," in his view, one that wanted to "create instability" in Central America and, in spite of improving conditions in the Eastern European satellites, one that required the country "to be prepared for the unexpected."[104]

Others within the administration, such as Secretary of Commerce Robert Mosbacher, apparently encouraged by the rapidly changing political conditions in Eastern Europe and the Soviet Union, and motivated by a strong desire to improve the nation's overseas trade position, worked for the relaxation of export restraints. Mosbacher's intentions in this regard became evident in July 1989 when, to Cheney's great displeasure, he relaxed export controls on some of the most popular brands of personal computers.[105] Broader export policy reform would follow.

An impetus for the administration to formulate its own reform proposals came with the reauthorization of the Export Administration Act, due to expire automatically at the end of September 1990. In early May, the House Committee on Foreign Affairs completed action on a bill extending the export statute for one year and making several major policy changes. The Bush Administration, through a letter to the committee chairman from national security adviser Brent Scowcroft, strongly objected to the measure and hinted it might be vetoed if presented to the President.[106]

Just before committee action on the reform bill was completed, the White House, announcing the conclusion of a comprehensive study of export controls on dual use technologies and goods, which was prepared under the direction of the Joint Chiefs of Staff and the federal intelligence community, released a major proposal for modernizing COCOM export arrangements. COCOM or the Coordinating Committee for Multilateral Export Controls, essentially consisting of the United States, its NATO allies, Australia, and Japan, vetted the transfer of high technology from its member nations to Communist bloc countries. The Bush Administration proposal recommended the creation of a new, superseding, and limited core list of controlled key technologies and goods, elimination of certain COCOM industrial list entries during the transition to the core list, establishment of parity for exports to Eastern Europe and the Soviet Union with those currently permissible to the People's Republic of China, and designation of three priority sectors for immediate or partial COCOM export decontrol–computers, telecommunications, and machine tools.[107]

Presenting this proposal at a high-level COCOM meeting in early

June, the administration supplemented it with a concession to allow the sale of advanced telecommunications equipment to newly democratic nations in Eastern Europe.[108] As this submission was being accepted in Paris, the House of Representatives approved the export reform bill on a 312–86 vote.[109] The Senate completed action on its amendments to the Export Administration Act in mid-September.[110] At the end of the month, with no reauthorization having been approved, President Bush, invoking national emergency powers, continued the Export Administration Regulations as the statute expired.[111] Congressional conferees reached agreement on the reauthorization legislation in late October[112] and both Houses gave approval to the conference version on October 26.[113] Ignoring congressional pleas to the contrary,[114] President Bush, however, withheld his approval of the measure and, with the 101st Congress having adjourned, it was "pocket vetoed."[115]

OVERVIEW

"The American business community," observed former Under Secretary of Commerce Lionel Olmer in his statement at a March 1990 House subcommittee hearing on Export Administration Act reauthorization, "has been the primary instigator of the search for change [in export policy], impelled by increasingly tough international competition and the costly time-consuming, frustrating and often unpredictable nature of export controls in their daily operation."[116] As this statement suggests, American business has pressed for relief from stringent export controls and, at the outset of the final decade of this century, appears to be obtaining it.

Campus-based science also enjoys some degree of renewed freedom regarding its professional communication, not because national security controls have been eliminated or are considered no longer applicable, but because they have been relaxed. They have been relaxed due largely to effective lobbying by American business, aided by a widespread desire among policymakers to increase the competitive position of the nation in international trade, and the end of the Cold War.

However, national security controls on traditional scientific communication may once again be increased for new reasons, such as promoting American competitiveness, preventing the proliferation of chemical and biological weapons, or combating terrorism. All of these considerations are within current rethinking of the national security concept. One of the more significant arenas where such rethinking has been occurring is the Office of Science and Technology Policy.[117]

Perhaps it will become the vehicle for realizing the policy balance sought by campus-based scientists regarding their professional communication in the uncertain world of the next century. Perhaps the end of the Cold War will bring such balance. Still, the control authority remains available, ready to be applied in new contexts of national security.

NOTES AND REFERENCES

[1]See 64 Stat. 149; also see J. Merton England. *A Patron for Pure Science: The National Science Foundation's Formative Years, 1945–1957.* Washington, DC: National Science Foundation, 1982.

[2]See 61 Stat. 495.

[3]See 63 Stat. 578.

[4]On the activities and accomplishments of OSRD, see: E.C. Andrus, Detlev W. Bronk, et. al. *Advances in Military Medicine.* Boston: Little, Brown and Company, 1948; James Phinney Baxter III. *Scientists Against Time.* Boston: Little, Brown and Company, 1946; Joseph C. Boyce, ed. *New Weapons for Air Warfare.* Boston: Little, Brown and Company, 1947; Charles W. Bray. *Psychology and Military Proficiency: A History of the Applied Psychology Panel of the N.D.R.C.* Princeton, NJ: Princeton University Press, 1948; John E. Burchard, ed. *Rockets, Guns and Targets.* Boston: Little, Brown and Company, 1948; William A. Noyes, Jr., ed. *Chemistry: A History of the Chemistry Components of the N.D.R.C., 1940–1946.* Boston: Little, Brown and Company, 1948; Irwin Stewart. *Organizing Scientific Research for War: The Administrative History of the Office of Scientific Research and Development.* Boston: Little, Brown and Company, 1948; Chauncy G. Suits, Louis Jordan, and George R. Harrison, eds. *Applied Physics: Electronics, Optics, Metallurgy.* Boston: Little, Brown and Company, 1948; Lincoln R. Thiesmeyer and John E. Burchard. *Combat Scientists.* Boston: Little, Brown and Company, 1947.

[5]U.S. Congress. House. Committee on Science and Technology. *Science Support by the Department of Defense,* p. 134.

[6]*Ibid.*, p. 133.

[7]*Ibid.*, p. 135.

[8]"Campus Unrest Affects Chemical Education." *Chemical and Engineering News,* v. 48, September 7, 1970, p. 38.

[9]See, generally, Dorothy Nelkin. *The University and Military Research: Moral Politics at MIT.* Ithaca, NY: Cornell University Press, 1972.

[10]U.S. Congress. House. Committee on Science and Technology. *Science Support by the Department of Defense.*

[11]U.S. Office of Science and Technology Policy. *Report of the Working Group on Basic Research in the Department of Defense.* June 22, 1978. Washington, DC, p. 11.

[12]See David Dickson. *The New Politics of Science.* New York: Pantheon Books, 1984, p. 128.

[13]See Nicholas Wade. "Science Meetings Catch the U.S.-Soviet Chill." *Science,* v. 207, March 7, 1980, pp. 1056, 1058; U.S. Congress. House. Committee on Government Operations. *The Government's Classification of Private Ideas.* H. Rept. 96–1540, 96th Congress, 2d Session. Washington, DC: U.S. Govt. Print. Off., 1980, pp. 103–104; G.B.L. "Government Bars Soviets from AVS and OSA Meetings." *Physics Today,* v. 33, April, 1980, pp. 81–83.

[14]Constance Holden. "Feds Defend Bubble Meddle." *Science*, v. 208, May 9, 1980, p. 577.

[15]See *Ibid.*; G.B.L., "Government Bars Soviets from AVS and OSA Meetings"; Wade, "Science Meetings Catch the U.S.-Soviet Chill."

[16]Gina Bari Kolata. "Attempts to Safeguard Technology Draw Fire." *Science*, v. 212, May 1, 1981, pp. 523–524, 526.

[17]*Ibid.*

[18]See *Ibid.*; Associated Press. "Restrictions on Export of University Research Attacked." *Washington Post*, April 9, 1981, p. A4. The full text of the presidents' letter, which was leaked to the press, appears in National Academy of Sciences. *Scientific Communication and National Security.* Washington, DC: National Academy Press, 1982, pp. 136–139.

[19]Barbara Crossette. "State Dept. Is Asking Colleges To Curb Research by Chinese." *New York Times*, November 27, 1981, pp. A1, A15.

[20]"From the President: On a Matter of Importance—Technical Information Flow." *Computer*, v. 15, May, 1982, p. 7.

[21]See Jeffrey R. Smith. "The Mysterious Chinese Luggage Incident." *Science*, v. 216, June 11, 1982, pp. 1204–1205.

[22]See Eliot Marshall. "Medical Data Bank: A Security Risk?" *Science*, v. 216, May 21, 1982, p. 831.

[23]See Philip M. Boffey. "Censorship Action Angers Scientists." *New York Times*, September 5, 1982, p. 1, 16; Ivars Peterson, Janet Raloff, and Joel Greenberg. "'Remote Censoring:' DOD Blocks Symposium Papers." *Science News*, v. 122, September 4, 1982, pp. 148–149; Gina Bari Kolata. "Export Control Threat Disrupts Meeting." *Science*, v. 217, September 24, 1982, pp. 1233–1234.

[24] U.S. Congress. House. Committee on the Judiciary. *1984: Civil Liberties and the National Security State.* Hearings, 98th Congress, 1st and 2d Sessions. Washington, DC: U.S. Govt. Print. Off., 1984, pp. 113–114.

[25] *Ibid.*, pp. 111–112.

[26]See Ross Gelbspan. "One Case of Papers Being Withdrawn." *Boston Globe*, January 23, 1984, p. 7.

[27]Colin Norman. "To Catch a Spy." *Science*, v. 222, November 25, 1983, pp. 904–905.

[28]See Philip M. Boffey. "3 Major Research Centers Reject Censorship." *New York Times*, April 10, 1984, p. A15; Colin Norman. "Universities Gag on Research Controls." *Science*, v. 224, April 13, 1984, p. 134.

[29]Colin Norman. "DeLauer Questions DOD Censorship." *Science*, v. 224, May 4, 1984, p. 471.

[30]U.S. Congress. House. Committee on Science and Technology. *Scientific Communications and National Security.* Hearing, 98th Congress, 2d Session. Washington, DC: U.S. Govt. Print. Off., 1984, p. 157; the draft policy statement appears in *Ibid.* at p. 154; also see John Walsh. "DOD Springs Surprise on Secrecy Rules." *Science*, v. 224, June 8, 1984, p. 1081.

[31]See Colin Norman. "Universities Prevail on Secrecy." *Science*, v. 226, October 26, 1984, p. 418.

[32]See Eliot Marshall. "Do Seminars Leak Navy Secrets?" *Science*, v. 224, June 29, 1984, p. 1409.

[33]Robert L. Park. "Intimidation Leads to Self-Censorship in Science." *Bulletin of the Atomic Scientists*, v. 41, March, 1985, p. 23.

[34]"AIAA Feels Chilling Effect of DOD Censors." *Defense Week*, v. 5, May 7, 1984, p. 6.

[35]Park, "Intimidation Leads to Self-Censorship in Science," p. 23.

[36]At the same time, some scientists and engineers were thought to be "bowing out of presentations at conferences just to avoid trouble, not because there is any real transfer problem." William H. Gregory. "The Technology Transfer Mess." *Aviation Week and Space*

Technology, v. 120, May 14, 1984, p. 13.

[37]There was also a countermovement, not only resisting self-censorship, but also reinforcing the practice of presumptively open symposia and conferences. In July 1985, the American Institute of Physics, for example, avowed it would not hold restricted sessions at any of its meetings in order that unclassified presentations deemed sensitive by the Pentagon might be made. See Colin Norman. "Physics Institute Will Not Restrict Meetings." *Science*, v. 229, July 19, 1985, p. 253.

[38]Robert L. Park. "The Muzzling of American Science." *Academe*, v. 72, September-October, 1986, p. 20.

[39]The authority in question was 10 U.S.C. 140c, subsequently recodified as 10 U.S.C. 130, concerning the protection of technical data by the Secretary of Defense; See David Burnham. "Pentagon Acts to Curb Science Parley Papers." *New York Times*, April 8, 1985, p. A15; David Burnham. "Pentagon Plan to Control Scientific Data Disputed." *New York Times*, April 9, 1985, p. A21; Colin Norman. "Security Problems Plague Scientific Meeting." *Science*, v. 228, April 26, 1985, pp. 471–472.

[40]Colin Norman. "Supercomputer Restrictions Pose Problems for NSF, Universities." *Science*, v. 229, July 12, 1985, p. 148.

[41]Colin Norman. "Illinois, Cornell Sign Supercomputer Contracts." *Science*, v. 229, August 9, 1985, p. 538.

[42]Philip M. Boffey. "Biotechnology May be Used Against U.S., Official Says." *New York Times*, May 28, 1984, p. 9; Philip J. Hilts. "Soviets May Be Using Biotechnology Advances in Weaponry, DIA Aide Says." *Washington Post*, May 28, 1984,p. A2.

[43]See Marjorie Sun. "U.S. Wants to Keep Eye on Biotech Exports." *Science*, v. 229, July 26, 1985, p. 368; David Silverberg. "Critical Technologies to Remain Under Export Controls." *Defense News*, v. 5, July 16, 1990, p. 14.

[44]See Colin Norman. "Memo Sets Policy for 'Star Wars' Publications." *Science*, v. 229, August 30, 1985, p. 843.

[45]Emphasis in original; among those signing the letter were the presidents of the American Association of Engineering Societies, the American Association for the Advancement of Science, the Institute of Electrical and Electronics Engineers, the National Society of Professional Engineers, the American Physical Society, the Society of Photo-Optical Instrumentation Engineers, the Society of Manufacturing Engineers, the Optical Society of America, the American Chemical Society, the American Society of Civil Engineers, the Institute of Industrial Engineers, and the System Safety Society.

[46]Responding to the letter of the 12 presidents refusing to sponsor closed or restricted sessions at meetings of their professional societies, Colonel Donald Carter, Acting Deputy Under Secretary of Defense for Research and Engineering reportedly indicated in early October 1985 that this action left the Pentagon with little choice but to sponsor separate, adjunct sessions at the same time and at the same place as a general scientific conference. This would enable regulation of topics discussed and attendees at such gatherings. R. Jeffrey Smith. "White House Issues Secrecy Guideline." *Science*, v. 230, October 11, 1985, p. 152.

[47]See Michael Schrage. "U.S. May Tighten Electronic Net to Control Software." *Washington Post*, May 6, 1984, pp. D1, D3.

[48]See, for example, John Horgan. "Air Force Proposal to Restrict Databases Faces Stiff Opposition." *The Institute*, v. 10, August, 1986, pp. 1–2.

[49]101 Stat. 1724.

[50]An unclassified version of NSDD 145 is reproduced here in the appendix of documents.

[51]Robert D. McFadden. "F.B.I. in New York Asks Librarians' Aid In Reporting on Spies." *New York Times*, September 18, 1987, pp. A1, B2.

[52]See Bill McAllister. "Librarians Want FBI to Shelve Requests About Foreign Readers."

Washington Post, March 27, 1988, p. A3; Natalie Robins. "The F.B.I.'s Invasion of Libraries." *The Nation*, v. 246, April 9, 1988, pp. 481, 498–502.

[53]U.S. Congress. House. Committee on the Judiciary. *FBI Counterintelligence Visits to Libraries*. Hearings, 100th Congress, 2d Session. Washington, DC: U.S. Govt. Print. Off., 1989; also see Herbert N. Foerstel. *Surveillance in the Stacks: The FBI's Library Awareness Program*. New York: Greenwood Press, 1991.

[54]See Harold C. Relyea, ed. *Striking A Balance: National Security and Scientific Freedom—First Discussions*. Washington, DC: American Association for the Advancement of Science, 1985, pp. 51–55; also see Bobby R. Inman/William D. Carey. "Classifying Science: A Government Proposal . . . And a Scientist's Objection." *Aviation Week and Space Technology*, v. 116, February 8, 1982, pp. 10–11, 82.

[55]Established in 1957 shortly after the Soviet Union had launched its Sputnik spacecraft, the Defense Science Board provides independent, expert advice to the Secretary of Defense, the chairman of the Joint Chiefs of Staff, and the Under Secretary of Defense for Acquisition. It is comprised of 31 at-large members, three ex officio members, and nine senior consultants. Strictly an advisory body, the Board conducts its activities through ad hoc task forces, usually comprised of at least two DSB members and additional independent experts in the field of interest to the particular task force. Reports produced by task forces are subject to Board approval before being transmitted to the Secretary and Under Secretary. The independence and objectivity of the DSB, and other similar military advisory panels, came into question by congressional overseers in 1983. See U.S. Congress. House. Committee on Government Operations. *Favoritism and Bias Within the Defense Science Board and Other Military Advisory Panels*. Hearing, 98th Congress, 1st Session. Washington, DC: U.S. Govt. Print. Off., 1984; U.S. Congress. House. Committee on Government Operations. *Defense Science Boards: A Question of Integrity*. H. Rept. 98–580, 98th Congress, 1st Session. Washington, DC: U.S. Govt. Print. Off., 1983.

[56]Almost identical commentary was offered by George P. Millburn, Acting Deputy Under Secretary of Defense for Research and Engineering, at a March 1982 hearing held by two subcommittees of the House Committee on Science and Technology. See U.S. Congress. House. Committee on Science and Technology. *Impact of National Security Considerations on Science and Technology*. Hearing, 97th Congress, 2d Session. Washington, DC: U.S. Govt. Print. Off., 1982, pp. 17, 25–26. Also see text accompanying note 87 of Chapter 2.

[57]U.S. Department of Defense. Office of the Under Secretary of Defense for Research and Engineering. *Report of the Defense Science Board Task Force on University Responsiveness to National Security Requirements*. Washington, DC: Department of Defense, 1982, p. 4-3. The unusual pagination in this report is somewhat confusing; the first number indicates the chapter in which the page appears and the second is the sequential positioning of the particular page within a chapter.

[58]*Ibid*.

[59]*Ibid*., pp. 4-6–4-7.

[60]*Ibid*., p. 5-4.

[61]*Ibid*., p. 6-2.

[62]On the chartering of executive branch advisory committees, see the Federal Advisory Committee Act, 5 U.S.C. App.

[63]The quotation indicates the study panel's general mission, which was further detailed regarding four major elements; see National Academy of Sciences. *Scientific Communication and National Security*. Washington, DC: National Academy Press, 1982, pp. ix-x.

[64]As identified in the panel's final report, the members were as follows:

Dale R. Corson, Chairman—President emeritus, Cornell University

Richard C. Atkinson—Chancellor, University of California, San Diego; former Director, National Science Foundation

John M. Deutsch—Dean of Science, Massachusetts Institute of Technology; former Under Secretary, Department of Energy

Robert H. Dicke—Einstein Professor of Physics, Princeton University; former member, National Science Board

Edward L. Ginzton—Chairman of the Board, Varian Associates

Mary L. Good—Vice President and Director of Research, UOP, Inc; member, National Science Board

Norman Hackerman—President, Rice University; former Chairman, National Science Board; former member, President's Science Advisory Committee

James R. Killian—President emeritus, Massachusetts Institute of Technology; former presidential science advisor; former member, President's Science Advisory Committee

Franklin Lindsay—Chairman, Executive Committee, ITEK Corporation

Richard A. Meserve—attorney, Covington and Burling

Wolfgang K.H. Panofsky—Director, Stanford Linear Accelerator Center, Stanford University; former member, President's Science Advisory Committee

William J. Perry—partner, Hambrecht and Quist; former Under Secretary for Research and Engineering, Department of Defense

Samuel C. Phillips—Vice President and General Manager, TRW Energy Products Group; former Director, National Security Agency

Alexander Rich—Sedgwick Professor of Biophysics, Massachusetts Institute of Technology; former member, National Science Board

John D. Roberts—Provost, California Institute of Technology; member of the Council, National Academy of Sciences

Harold T. Shapiro—President, University of Michigan

Charles P. Slichter—professor of physics, University of Illinois; former member, President's Science Advisory Committee; member, National Science Board

Michael I. Sovern—President, Columbia University

Elmer B. Staats—former Comptroller General of the United States; former Deputy Director, Bureau of the Budget; former Executive Officer, Operations Coordinating Board

[65]National Academy of Sciences. *Scientific Communication and National Security*, pp. x-xi.

[66]Contributors, briefers, and liaison representatives, an annotated bibliography, commissioned studies, working papers, and other related materials were appended to the panel's final report, which was initially produced in a two-volume press version, accompanied by various pertinent papers, and later published in a single book of approximately 190 pages.

[67]See Philip M. Boffey. "Security of U.S. Said to Be Hurt By Data Leaks." *New York Times*, October 1, 1982, pp. A1, A11; Philip J. Hilts. "Panel Discounts Scientific Security Leaks." *Washington Post*, October 1, 1982, p. A2; Gina Bari Kolata. "Security Controls Hurt Research, NAS Warns." *Science*, v. 218, October 15, 1982, pp. 271-272; Arlen J. Large. "Science Panel Says U.S. Shouldn't Classify Research Unless the Soviets Would Benefit." *Wall Street Journal*, October 1, 1982, p. 37.

[68]See Rosemary Chalk. "Commentary on the NAS Report." *Science, Technology, and Human Values*, v. 8, Winter, 1983, pp. 21-24.

[69]National Academy of Sciences. *Scientific Communication and National Security*, p. 1.

[70]*Ibid.*, p. 45.

[71]*Ibid.*, p. 48.

[72]*Ibid.*, p. 53.

[73]*Ibid.*, pp. 39-40.

[74]*Ibid.*, pp. 65-70; for details on recommendations, see *Ibid.*, pp. 39-64; the panel's four criteria are specified in the text accompanying note 50 in Chapter 6.

[75]Philip J. Hilts, "Panel Discounts Scientific Security Leaks," p. A2.

[76]Author's notes, National Academy of Sciences press conference, September 30, 1982.

[77]Hedrick Smith. *The Power Game: How Washington Works.* New York: Random House, 1988, p. 603.

[78]See "Hardliners Stonewalled NAS Secrecy Study." *Science & Government Report*, v. 12, October 15, 1982, pp. 1–3.

[79]NSSDs, like the National Security Decision Directives they frequently generate, are usually security classified and otherwise unpublished. The text of NSSD 14–82, which was not classified, is reproduced here in the appendix of documents. For background on such national security planning and policy instruments, see Harold C. Relyea. "The Coming of Secret Law." *Government Information Quarterly*, v. 5, no. 2, pp. 97–116.

[80]Philip M. Boffey, "Security of U.S. Said to Be Hurt By Data Links," p. A11.

[81]See Colin Norman. "Musical Chairs at OSTP." *Science*, v. 220, June 17, 1983, p. 1255.

[82]Mitchel B. Wallerstein and Lawrence E. McCray. "Scientific Communication and National Security: The Issues in 1984." January, 1984. 31 p.

[83]See Mitchel B. Wallerstein. "Scientific Communication and National Security in 1984." *Science*, v. 224, May 4, 1984, pp. 460–466.

[84]Quoted language is drawn from the initial January evaluation prepared for COSEPUP members and varies somewhat stylistically, but not substantively, from the *Science* magazine text.

[85]See U.S. Congress. House. Committee on Science and Technology. *Scientific Communications and National Security.* Hearing, 98th Congress, 2d Session. Washington, DC: U.S. Govt. Print. Off., 1984, pp. 137, 154–160; the draft NSDD appears on p. 154.

[86]*Ibid.*, pp. 137–138.

[87]See *Ibid.*, pp. 156–157.

[88]The text of NSDD 189, which was not classified, is reproduced here in the appendix of documents.

[89]See U.S. Congress. House. Committee on Science and Technology. *Scientific Communications and National Security*, pp. 48–52.

[90]See U.S. Congress. Committee on Government Operations. *The Government's Classification of Private Ideas.* Hearings, 96th Congress, 2d Session. Washington, DC: U.S. Govt. Print. Off., 1981; U.S. Congress. House. Committee on Government Operations. *The Government's Classification of Private Ideas.* H. Rept. 96–1540, 96th Congress, 2d Session. Washington, DC: U.S. Govt. Print. Off., 1980.

[91]See, for example, U.S. Congress. House. Committee on Science and Technology. *Impact of National Security Considerations on Science and Technology.* Hearing, 97th Congress, 2d Session. Washington, DC: U.S. Govt. Print. Off., 1982; U.S. Congress. House. Committee on Science and Technology. *Scientific Communications and National Security.*

[92]See U.S. Congress. House. Committee on the Judiciary. *1984: Civil Liberties and the National Security State.* Hearings, 98th Congress, 1st and 2d Sessions. Washington, DC: U.S. Govt. Print. Off., 1984.

[93]99 Stat. 120.

[94]CSIS Science and Technology Committee. Panel on Technology Transfer. *Securing Technological Advantage: Balancing Export Controls and Innovation.* Stephen A. Merrill, ed. Washington, DC: Center for Strategic and International Studies, 1985.

[95]Business-Higher Education Forum. *Export Controls: The Need to Balance National Objectives.* Washington, DC: Business-Higher Education Forum, 1986.

[96]National Academy of Sciences. *Balancing the National Interest.* Washington, DC: National Academy Press, 1987; also see National Academy of Sciences. *Balancing the National Interest: Working Papers.* Washington, DC: National Academy Press, 1987.

[97]National Research Council. *Global Trends in Computer Technology and Their Impact on Export Control.* Washington, DC: National Academy Press, 1988; also see Seymour E. Goodman, Marjory S. Blumenthal, and Gary S. Geipel. "Export Control Reconsidered." *Issues in Science and Technology*, v. 6, Winter, 1989–1990, pp. 58–62.

[98]National Academy of Sciences. *Finding Common Ground: U.S. Export Controls in a Changed Global Environment*. Washington, DC: National Academy Press, 1991.

[99]These studies are discussed in the next chapter on industry and new technologies.

[100]See 102 Stat. 1107, 1347–1370.

[101]102 Stat. 1363.

[102]See Molly Moore. "Cheney Predicts Gorbachev Will Fail, Be Replaced." *Washington Post*, April 29, 1989, p. A17; also see Patrick E. Tyler. "New Pentagon 'Guidance' Cites Soviet Threat in Third World." *Washington Post*, February 13, 1990, pp. A1, A9; R. Jeffrey Smith and Patrick E. Tyler. "Cheney, DIA Said to Be at Odds on Soviet Threat." *Washington Post*, March 8, 1990, p. A7; Molly Moore and Patrick E. Tyler. "Secretary Cheney: Still the Skeptic." *Washington Post*, March 21, 1990, pp. A1, A8; Patrick E. Tyler. "Cheney Again Predicts Failure For Gorbachev and Reforms." *Washington Post*, April 30, 1990, p. A17.

[103]Paul Bedard. "Quayle: Let Soviets Reform Themselves." *Washington Times*, October 18, 1989, pp. Al, A10; Morton Kondracke. "Quayle, Baker Square Off." *Washington Times*, October 31, 1989, pp. F1, F4.

[104]David S. Broder. "Quayle Skeptical of Soviets." *Washington Post*, December 5, 1989, pp. A1, A42.

[105]See Stuart Auerbach. "U.S. Relaxes Computer Sales Curbs." *Washington Post*, July 19, 1989, pp. A1, A11; Molly Moore and Ann Devroy. "Cheney Assails Relaxation Of Computer Sales Controls." *Washington Post*, July 20, 1989, p. A10; David Silverberg. "Commerce Wins Its Latest Trade Turf Battle With Pentagon." *Defense News*, v. 4, July 24, 1989, p. 15.

[106]U.S. Congress. House. Committee on Foreign Affairs. *The Reauthorization of the Export Administration Act*. Hearings and markup, 101st Congress, 2d Session. Washington, DC: U.S. Govt. Print. Off., 1990; U.S. Congress. House. Committee on Foreign Affairs. *Export Facilitation Act of 1990*. H. Rept. 101–482, 101st Congress, 2d Session. Washington, DC: U.S. Govt. Print. Off., 1990.

[107]Ann Devroy. "U.S. Seeks To Relax Trade Curbs." *Washington Post*, May 3, 1990, pp. A1, A39; David Silverberg. "White House Export Control Initiative Meets Mixed Reviews." *Defense News*, v. 5, May 7, 1990, p. 6.

[108]Stuart Auerbach. "U.S. Backs Sales to East Bloc." *Washington Post*, May 24, 1990, pp. E1, E5; Stuart Auerbach. "High-Tech Sales Plan Unveiled." *Washington Post*, June 7, 1990, pp. E1, E4; Stuart Auerbach. "High-Tech Trade Curbs Eased for Eastern Europe." *Washington Post*, June 8, 1990, pp. Fl-F2; Stuart Auerbach. "U.S. Relaxes Licensing Of High-Tech Exports." *Washington Post*, June 19, 1990, pp. D1, D3; Giovanni deBriganti. "Both Sides Bend at CoCom." *Defense News*, v. 5, June 11, 1990, pp. 4, 26; Theresa Hitchens and David Silverberg. "CoCom To Create New Era In Trade." *Defense News*, v. 5, June 4, 1990, pp. 1, 68

[109]See *Congressional Record*, v. 136, June 6, 1990, pp. H3270-H3355.

[110]See *Congressional Record*, v. 136, September 13, 1990, pp. S13004-S13041.

[111]See E.O. 12730, *Federal Register*, v. 55, October 2, 1990, pp. 40373–40374.

[112]See *Congressional Record*, v. 136, October 25, 1990, pp. H13844-H13871.

[113]See *Congressional Record*, v. 136, October 26, 1990, pp. H12771-H12783, S17179-S17191.

[114]David Silverberg. "Senators Urge Bush to Sign Export Law." *Defense News*, v. 5, November 5, 1990, p. 12.

[115]For the President's objection to the bill, see *Weekly Compilation of Presidential Documents*, v. 26, November 19, 1990, pp. 1839–1840.

[116]U.S. Congress. House. Committee on Foreign Affairs. *The Reauthorization of the Export Administration Act*, p. 399.

[117]Wil Lepkowski. "Bromley at Midterm Is Confident, In Charge In Science Adviser Job." *Chemical and Engineering News*, v. 69, May 20, 1991, pp. 21–23.

5

THE INDUSTRY SETTING

Apart from the university setting, scientific and technological research came to be performed in the United States in at least two other environments – independent invention and industry. The campus tradition was ideally one of *pure science*, conducting studies for the purpose of making new discoveries and adding to the store of knowledge, unconcerned about commercial advantage. Research results were published to establish their originality, priority, and authenticity, with little resort to formalities of intellectual property protection. This tradition arose from the practice of the towering 19th-century British scientist Michael Faraday, who refused to accept any economic profit from the commercial application of his discoveries.[1] During the latter half of the present century, however, research other than pure science came to be performed in the university setting.[2]

Scientific and technological research was also conducted in America by independent inventors – men such as Alexander Graham Bell, Thomas A. Edison, and Elmer Sperry. Their primary era ranged from the final quarter of the 19th century through the years of World War I, beginning, by one assessment, about the time Bell invented his telephone and Edison established his laboratory in Menlo Park, New Jersey.[3] Their accomplishment was nothing less than to establish the United States, by the eve of the Great War, as "the most inventive of all nations."[4] Four years later, however, as hostilities in Europe concluded, the independent inventors began to be eclipsed and displaced by another research group – industrial scientists.

The independent inventors, as scientists, were practical problem solvers. Largely freed from organizational constraints, they performed research on matters that interested them, applying, in varying degrees, scientific methods and knowledge to effect greater compatibility between humankind and the environment or to give man more control over his surroundings. The results were "the nuclei of technological systems." Concentrating on the primary act of invention, they were interested in entrepreneurial activities largely as a means for bringing their inventions into use. Success was measured mainly in terms of the number of patents held, and the patents of others were a rich and closely followed source of ideas for both identifying problems and solving them.[5]

Unlike the academic scientists, the independent inventors did not produce isolated bits of knowledge explaining natural phenomena and they were not interested in abstract theory. For the most part, their research was concentrated within a particular area of study, yielding compounded results—a series of inventions utilizing a common resource, such as electrical energy or cybernetics, or refining a single technique, such as powered flight. They sought scientific knowledge and theory explaining the state of the art in their problem area, not what was beyond it.[6]

With the advent of world war, American military authorities supported the endeavors of the independent inventors and developed such large-scale technological systems as wireless telegraphy and radio, the airplane, and both nautical and aeronautical guidance-and-control mechanics. This experience revealed a basic problem in relying upon independent inventors: Most of them had no manufacturing facilities. Arrangements had to be made for American industries to mass produce and to fine tune the contributions of the independent inventors. Furthermore, the independent inventors tended to offer radically new devices rather than inventions improving and expanding existing systems. Consequently, American corporations met these needs with manufacturing capabilities and industrial research scientists to promote production. By the end of the war, industrial scientists, not independent inventors, were largely changing the material world, with a concomitant shift in American invention from a revolutionary to an evolutionary mode. Professional inventors would continue to make their independent, radical contributions during the century—such as air conditioning, power steering, cellophane, the jet engine, and the laser—but never in the quantity or with quite the same style as in the past.[7]

Initially established in Germany around the time of that nation's unification, industrial research laboratories appeared in America just after the turn of the century and gained prominence during World War I. There were at least 100 such facilities in the U.S. prior to 1914, and more than 1,000 of them at the onset of the Great Depression 15 years later. They were functionally conservative and largely content with improving existing technological systems and product lines. Recruits were bright, young, inventive, and practical-minded men with no entrepreneurial ambitions of their own. Moreover, they did not feel professionally inhibited by the directed research priorities of industry or by emerging trade secrecy safeguards. Professional communication, although understandably subject to some corporate control, was conducted at various levels, including teaching and public dissemination of research findings. For example, Charles Peek, Jr., a 1903 engineering graduate of Stanford University working at Charles Steinmetz's General Electric laboratory, produced more than 20 scholarly papers between 1911 and 1931.[8]

PRODUCTS, PROCESS, AND KNOWLEDGE

The industrial scientists applied and improved technologies not only for better goods and services, but also for improved production—new market offerings in greater numbers at less cost. Thus, the manufacturing process and production system became at least as important, if not more so, as what was engendered. In an export context, recognition of this situation was expressed in a 1976 Defense Science Board task force study directed by Texas Instruments executive vice president J. Fred Bucy: "the transfer of design and manufacturing know-how is of overwhelming importance to our national security. It is mastery of design and manufacturing that increases a nation's capability," said the Bucy report, "and it is in this area that the U.S. maintains its technological leadership."[9] Subsequently, the task force recommended that, "Three categories of export should receive primary emphasis in control efforts, since they transfer vital design and manufacturing know-how most effectively:

- Arrays of design and manufacturing information that include detailed 'how to' instructions on design and manufacturing processes.
- 'Keystone' manufacturing, inspection, or automatic test equipment.
- Products accompanied by sophisticated operation, application, or maintenance, information.[10]

This approach to technology transfer and the task force's proposed Militarily Critical Technologies List to identify the most militarily significant technologies, among other recommendations by the panel, were realized with the Export Administration Act of 1979.[11]

At first blush, it might appear that exporters in the business community would have had little objection to the new protection of vital design and manufacturing know-how as it seemingly would coincide with their own interests in protecting these same vital elements for entrepreneurial reasons. Similarly, it might have been thought that there would have been no quarrel with the Militarily Critical Technologies List that was mandated as a means of reviewing and assessing the adequacy of the Commodity Control List used by the Commerce Department to regulate exports and possibly to arrive at a more limited control list of only the most militarily significant technologies.

Protest of and objection to both of these innovations, however, did result. The new design and manufacturing know-how protections were interpreted too broadly so that, contrary to the warnings of the Bucy report,[12] business meetings, training agreements, technical exchanges, and workshops were affected.[13]

Furthermore, the Bucy report's insistence on depriving the East bloc of U.S. design and manufacturing know-how constituted control of information—a flow of information vital for American technology development. "No high technology firm, not even the grandest, can afford to be isolated from external sources of information," observed economist Stuart Macdonald, "especially customers, suppliers, and competitors."[14] Considering implementation, Lewis Branscomb, former chief scientist for the IBM Corporation now teaching at Harvard University, commented:

> There is no way a government agency can know enough about the dynamics of knowledge in fast-moving commercial technologies to select out those kernels of critical knowledge that might advantageously be restricted. The result is a system that threatens the effectiveness of the information flow system in our society without accomplishing much of value to our defense.[15]

The Militarily Critical Technologies List, a huge, security classified compilation, covers some 700 technologies and neither reduced nor was integrated into the Commodity Control List.[16] Moreover, the Pentagon gave serious consideration in 1983 to expanding it with an appended Militarily Significant Emerging Technologies Awareness List.[17] Commenting that the MCTL "is the size of the New York phone book and worth a lot less," Fred Bucy subsequently called it "a bastardization of the concept of critical technology" and blamed the Pentagon bureaucracy for making it "absolutely worthless."[18]

The MCTL "should be on five sheets of paper," he asserted.[19] Robert Schmidt, vice president of Control Data Corporation, characterized it as "an exercise in futility."[20] Furthermore, the determination that a dual use technology is militarily critical or on the MCTL is usually sufficient to subject it to export control. The Bucy report noted several factors for consideration when determining whether or not a technology can be effectively transferred, including ongoing interaction between the receiver and donor of the technology, and the extent to which the receiver has the following:

- A well-defined goal and adequate resources committed to accomplishing it.
- Key individuals competent in the technology, who will be directly involved in applying the newly received technology.
- An adequate infrastructure capable of providing necessary parts, supplies, instrumentation, and manufacturing equipment.[21]

Bucy himself offered another important consideration: "Controlling militarily critical technologies . . . requires a determination of an adversary's military capability and the potential effect of a technology on that capability."[22] In the case of the Soviet Union, some question existed concerning the extent to which that country had the ability to adopt, integrate, and use American technology.[23]

American export industries publicly supported the strict controls mandated by the Export Control Act of 1949,[24] and, through the 1960s, apparently accepted the restraints of the prevailing export laws.[25] Then, with the thaw in U.S.-Soviet relations and the onset of "peaceful coexistence," exporters in the business community and many members of Congress sought to liberalize trade with the Communist bloc. It had also become apparent that the Soviet Union could now acquire from other sources many industrial and strategic goods that were restricted by U.S. export control laws.[26] With the Export Administration Act of 1969,[27] Congress shifted from a policy of strategic embargo to one of qualified free trade with the East bloc.[28] Significant realization of expanded East-West trade was delayed by presidential opposition for about three years until détente was achieved. Industrial exporters were unhappy with the administration's initial lack of enthusiasm for expanded trade with the Communist bloc. As this situation changed, particularly with new diplomatic and trade relations with the People's Republic of China, exporters became increasingly frustrated with licensing delays, extreme cautiousness in allowing shipments of some goods, and the failure of the Commerce Department to establish a capacity for foreign availability assessment. These problems remained as détente collapsed and the nation elected Ronald Reagan.

The new Secretary of Defense, Caspar W. Weinberger, saw the country as being highly susceptible to Soviet coercion and aggression. In his view, the strategic position of the United States, defensively and offensively, was weak. Amassing greater amounts of more capable weapons and personnel alone, however, was not the only corrective. Other reforms, affecting foreign policy, domestic commerce, and international trade, would be pursued.

Weinberger's success in these endeavors was predicated on several important factors. He was a highly experienced public official—a veteran of the Nixon Administration who stayed on to assist President Gerald Ford during his first year in office. His reputation was one of integrity, and he was personally close to President Reagan. He probably found an ally in another high-ranking Reagan friend, William J. Casey, the new Director of Central Intelligence. Complaints and opposition from the Department of Commerce were limited by couching initiatives in "national security" policy terms. Later, "Defense did its utmost to discredit Commerce as the agency chiefly responsible for the administration and enforcement of export controls, and it did so as publicly as possible."[29]

The most likely source of challenge to Weinberger's assertiveness in external relations—Secretary of State Alexander M. Haig—self-destructed. A distinguished career Army officer and recently tested veteran of several administrations, Haig yielded to few in his hard line attitude toward the Soviet Union. However, he evidenced a "brittle, confrontational style," and his "connection to the President was new and thin."[30] Moreover, his desire to be the Reagan Administration's preeminent agent in foreign policy formulation and execution, which could have brought him into conflict with Weinberger, was his undoing. Between the election and the inauguration, he submitted a draft presidential memorandum to Reagan's principal aides. If signed, it would have given Haig unparalleled formal authority as Secretary of State over the foreign affairs community. Not only was the proposal rejected, but also, in response, Haig, Weinberger, and Casey were given coequal coordinating responsibility within their separate policy spheres, and a crisis management committee was created under the chairmanship of the Vice President, who, of course, was not a full-time foreign policy player.

In March 1981, while visiting Capitol Hill, Haig publicly criticized the crisis committee that the President was about to mandate. In response, a presidential press statement was quickly released announcing the creation of the panel, affirming the Vice President's leadership of it, and insisting that crisis management had traditionally and appropriately been a White House responsibility. No mention was made of the Secretary of State.

A few days after this humiliation, Haig further damaged himself with his nervous "I am in charge here" announcement from the White House in the hours immediately after the attempt on President Reagan's life. In his remaining 15 months, his stock continued to fall as he intermittently skirmished with Weinberger, the President's national security adviser, the ambassador to the United Nations, and the three principal White House assistants, among others.[31]

Consequently, "Caspar Weinberger was more active on foreign policy than any of his predecessors since Robert McNamara."[32] Moreover, he brought to the Pentagon men who were ready and able to support him in this regard, such as Fred Ikle, the Under Secretary for Policy, and Richard Perle, the Assistant Secretary for International Security Policy.[33] His strong influence continued through the second year of the administration after Haig's departure because the new Secretary of State, George Shultz, did not immediately involve himself with U.S.-Soviet relations and arms control.[34] Furthermore, one of the authorities Weinberger sought to bring to bear upon the East bloc, the Export Administration Act, was not within the State Department's jurisdiction.

Within the Reagan Administration, Weinberger and his colleagues fostered the view that "the Soviet economy was in crisis" and that the United States, departing from the position of recent administrations, "should capitalize on this situation" by pursuing "not a policy of economic diplomacy but one of economic warfare."[35] In fulfillment of this objective, Weinberger "took advantage of every opportunity to eliminate or reduce the supply of Western credits and technology to the Soviet Union." Among these efforts, "a major campaign was launched to curb unauthorized exports to the U.S.S.R. and to curb Soviet industrial espionage."[36] He gave serious consideration to a plan to assign military units or the National Guard to patrol the docks and halt shipments of so-called "high-technology secrets" to the Soviet Union.[37] By March 1984, he convinced the President to give the Pentagon authority to review—and, in the view of some, effectively veto—critical high technology exports to a dozen countries suspected of diverting such goods to the East bloc. This expanded mandate not only added new power to the Pentagon's export regulation arsenal, but also constituted a major jurisdiction victory for DOD over the Commerce Department.[38]

INADEQUACIES

Weinberger and his colleagues, including some allies outside of the Pentagon, like Lawrence Brady at the Commerce Department, began vigorously pursuing efforts to reduce the transfer of militarily critical technology to the Soviet Union and the East bloc.[39] However, the leakage was difficult to assess. During the 1970s, a time of liberalized U.S.-Soviet trade and detente, sensitive American technology was nonetheless illegally acquired through espionage, theft, shipment diversion, and illicit sales. The breakdown of detente may have increased efforts of this kind.[40]

Did laxity or greed on the part of U.S. industry and commercial exporters foster such leakage? Although these possibilities cannot be dismissed, there appear to have been other contributing conditions of greater consequence. Investigating journalists and Senate overseers, for example, found that enforcement resources available to the U.S. Customs Service at the Treasury Department and the Office of Export Administration at the Commerce Department were weak, if not inadequate, and that staff inexperience and jurisdiction disputes among the enforcement agencies also were impediments.[41] The General Accounting Office found existing export regulation to be wasting scarce resources:

> Of the 60,783 export applications that Commerce reviewed unilaterally in fiscal year 1981, almost half could have been eliminated from licensing requirements and controlled in a less burdensome way because the products involved are not considered militarily significant. Such a large number of low-technology applications makes the licensing system more a paper exercise than a control process and detracts from the seriousness with which controls should be viewed. Submitting these applications costs industry approximately $6.1 million a year and the Government $1 million a year in unnecessary administrative costs. Other potential savings could be made by exploring various options to reduce or eliminate high-technology license requirements for exports to U.S. allies without totally relinquishing control over such items.[42]

Leakage assessment was also complicated by Pentagon efforts to discredit the Commerce Department's ability to deal with the transfer problem and a certain amount of press complicity in portraying the issue.[43] Some stories—that the Soviets had somehow built their An-76 cargo transport from U.S. C-141 technology, that Apple II computers with standard software were militarily critical because they could be used to guide nuclear weapons, or that an environmental impact statement in a Tennessee public library contained sufficient information to enable Soviet engineers to reconstruct a weapons manufacturing plant—soon lost their credibility.[44] Other offerings are more problematical.

In mid-September 1985, Secretary Weinberger released a report that he called "deeply sobering."[45] Prepared by the Technology Transfer Intelligence Committee (TTIC)—an interagency panel chaired by the Central Intelligence Agency with representatives from 22 federal agencies, including the entire intelligence community—the unclassified white paper, devoid of any identification as to its origins, stated at the outset that "the United States Government has learned of a massive, well-organized campaign by the Soviet Union to acquire Western technology illegally and legally for its weapons and military equipment projects." The TTIC report sought "to reveal in detail the structure of these Soviet programs, and to give examples of Soviet requirements and successes."[46]

Two Soviet acquisition programs were described. Greater attention was given to the activities of the Military Industrial Commission (VPK) of the Presidium of the Council of Ministers. Characterized as "the most powerful organization in defense production" by the report:

> the VPK seeks one-of-a-kind military and dual-use hardware, blueprints, product samples, and test equipment to improve the technical levels and performance of Soviet weapons, military equipment, and defense manufacturing equipment and reduce any dependency on advanced Western products. This is done by exploiting and adapting design concepts embodied in acquired equipment and associated documents.[47]

Apart from industrial sources, commercial databases, professional and academic conferences on applied science and technology, and research universities were identified as particular targets of VPK acquisition efforts.[48]

In addition, the report indicated, "the Ministry of Foreign Trade and Soviet intelligence services administer a trade diversion program to acquire relatively large numbers of dual-use manufacturing and test equipment for direct use in production lines."[49]

For its sources, the TTIC white paper referred vaguely to "information . . . obtained directly by the United States and Allied countries" and to "numerous sources and data."[50] It was soon learned, however, that production of the report was made possible, in large part, by the so-called Farewell papers. Farewell was the code name for a Soviet Committee for State Security (KGB) officer stationed in Paris during the 1960s and 1970s who assisted French intelligence. This individual provided detailed information on Soviet science and technology acquisition organization and activities in 1981. After returning to the Soviet Union, the officer resumed contact with French intelligence for a while, then ceased altogether.[51]

Although it can be conceded that at the time the TTIC white paper was produced and released, the Soviet Union maintained a massive, well-financed, and frequently effective effort to acquire American technology, the report raised a number of disquieting questions and issues. Was Farewell to be trusted? Could he have been a double agent, a source of disinformation? Were Farewell's contributions about Soviet efforts to target and acquire Western technology reliable accounts by objective analysts or inflated reports of success by self-serving bureaucrats? Were the discussions of scientific conference and academic-related collection efforts attempts to offset the 1982 findings and recommendations of the Corson panel? Was the white paper's September 18 release designed to minimize or respond to the September 17 letter of 12 scientific and engineering society presidents refusing to hold closed or restricted access technical sessions at meetings or conferences sponsored by their organizations? Perhaps it was timed to temper the scientific, technical, and engineering information transfer policy announced with the issuance of NSDD 189 a few days later on September 27. Did the report describe the conditions that would justify resort to the "applicable U.S. statutes" mentioned as a caveat to the policy expressed in NSDD 189? Whatever the answers might be, the white paper contributed little toward building mutual understanding and trust among American scientists and federal officials concerning leakage.

The TTIC report also indicated growing government interest in regulating access to computer databases. "The individual abstracts or references in government and commercial databases are unclassified," the white paper acknowledged, "but some of the information, taken in the aggregate, may reveal sensitive information concerning U.S. strategic capabilities and vulnerabilities." Calling public and private scientific and technical information clearinghouses "fertile ground" for Soviet collection efforts, the report warned that "the growing use of electronic data bases has provided the Soviets with an even more efficient means of identifying and procuring such unclassified technical information needed by Soviet designers."[52]

Government concern about the leakage of sensitive technical information in electronic formats became apparent near the end of President Reagan's first term. In May 1984, it was reported that the administration was considering expanded electronic surveillance activity to prevent the smuggling of computer software and data, whether campus or commercially generated, through international telephone contact.[53] Then, during the summer of 1986, an Air Force proposal to prevent direct access by noncitizens to unclassified U.S. electronic databases drew sharp criticism. Both government and private databases were to be so restricted, including commercial services such as Lockheed Corporation's Dialog, which links campus, research, and corporate systems.

Sensitive, but unclassified information—such as the results from research sponsored by the Pentagon—would be removed from the open databases and placed in ones accessible only to U.S. citizens. Benjamin Leon, chairman of the Technology Transfer Committee of the Institute of Electrical and Electronics Engineers objected: "Any restrictions placed on access to these databases would impede the ability of the United States to advance both military and civilian technology." Furthermore, he observed, the special databases with sensitive information could become attractive targets for East bloc agents who would likely marshall substantial resources to penetrate them.[54]

The Reagan Administration, however, was already perfecting a database restriction policy. The effort was begun with NSDD 145 of September 17, 1984. This unpublished, never fully declassified presidential directive purported to prescribe national policy on telecommunications and automated information systems security.[55] It gave the National Security Agency authority for prescribing computer and telecommunications security standards.[56]

Then, in November 1986, the reach of the administration's new policy became apparent. Speaking at a meeting of the Information Industry Association in New York City, Diane Fountain, Director of Information Systems, Office of the Assistant Secretary of Defense for Command, Control, Communications, and Intelligence told her audience: "I don't believe the issue is whether or not we are going to protect information." Instead, she told her startled listeners, "I really believe that the issue is what information we are going to protect, both from the federal government, both within DOD and also within industry."[57] Then a policy statement, issued pursuant to NSDD 145, was unveiled. This was National Telecommunications and Information Systems Security Policy Number 2 of October 29, 1986, signed by presidential national security adviser John M. Poindexter. It defined the information that the Pentagon, through various telecommunications and automated information systems security techniques, sought to protect in terms of dissemination, saying:

> Sensitive, but unclassified information is information the disclosure, loss, misuse, alteration, or destruction of which could adversely affect national security or other Federal Government interests. National security interests are those unclassified matters that relate to the national defense or the foreign relations of the U.S. Government. Other government interests are those related, but not limited to the wide range of government or government-derived economic, human, financial, industrial, agricultural, technological, and law enforcement information, as well as the privacy or confidentiality of personal or commercial proprietary information provided to the U.S. Government by its citizens.[58]

The immediate implication of the directive was that federal agencies could withhold sensitive information from commercial data vendors unless there was a commitment by them to maintain such information in databases accessible only to U.S. citizens. Some feared more grandiose efforts at restriction.[59] Many in Congress came to regard NSDD 145 a dangerous and unwarranted expression of policy that encroached upon the legislature's constitutionally prescribed lawmaking authority while assigning regulatory power to a quasi-military agency that more properly should be exercised by a civilian entity, and otherwise assumed a need to control "sensitive, but unclassified" information held in private sector databanks. NTISSP No. 2 was particularly odious not only because it purported to be legitimate as a consequence of being signed by an unelected presidential adviser, but also because it was quickly becoming apparent that the particular signatory was deeply involved in the emerging Iran-Contra affair.[60] Consequently, NSDD 145 was largely displaced by the Computer Security Act of 1987, which gave the National Bureau of Standards, now the National Institute of Standards and Technology, responsibility "for developing standards and guidelines for Federal computer systems, including responsibility for developing standards and guidelines needed to assure the cost-effective security and privacy of sensitive information in Federal computer systems."[61] There have been some difficulties with the implementation of the statute, not the least of which has been NIST dependency upon National Security Agency resources,[62] and security requirements for private sector computers were a special challenge.[63]

ASSESSMENTS AND REFORMS

The controversy over increased national security controls on scientific communication and technology transfer subsided somewhat in the mid-1980s. There were several reasons for this lull. Further modification of the Export Administration Act was effectively precluded in July 1985 when Congress reauthorized the statute for four years. The issuance of NSDD 189 in September convinced some in the scientific community that the government would be more restrained in its regulation of traditional scientific communication. Self-censorship at scientific professional conferences and symposia contributed to this false sense of relief. The public was not yet aware of NSDD 145 and its policy significance for private sector databases.

Nonetheless, at the time of this relative quiescence, there were some important indications of awareness that the general problem had not been resolved. Serious studies of the U.S. and multilateral national security export controls were undertaken by several entities. The inclusion of both university scientists and industrial leaders on the managerial panels for these studies suggested that perhaps economics was making business an ally in campus-based science's fight against the attempts of the federal government to stem the free flow of ideas. Such awareness had been expressed a few years earlier by Tom Christiansen, Hewlett-Packard Company's very knowledgeable manager of international trade relations, who would serve on the National Academy of Sciences' 1987 study committee following up on the Corson panel. "There has been a dual standard," he said. "Universities have had relatively little Government control of information, while companies have had a great deal." However, worried about the new effort to tighten information controls on universities for reasons of national security, Christiansen commented: "It would constrict the free flow of information between companies and universities: We'd be shooting ourselves in the foot."[64]

A report on one such study was published in the summer of 1985 by the Center for Strategic and International Studies in Washington, DC. Prepared by the Panel on Technology Transfer of the CSIS Science and Technology Committee, *Securing Technological Advantage: Balancing Export Controls and Innovation* found the contemporary technology transfer setting objectionable for reasons of both policy content and resulting political adversity.[65] "The Reagan administration has not supported a return to indiscriminate trade denial," observed the report, "but neither has it defined the limits of its broader conception of strategic technology or its effort to subject various transfer mechanisms to closer scrutiny and stricter control."[66] Officials and other observers contending "that control efforts in the early 1970s had been too lax to contain the Soviet threat" had not subjected their tougher stance to "a direct debate of opposing policies and assumptions," and Congress and the executive branch had battled to an impasse after a two-year effort to reauthorize the Export Administration Act during the 98th Congress.[67] The competitive position of American business and the nation in international trade was not helped by the resulting situation. According to its report, "the Panel believes that the U.S. approach to national security export controls exhibits serious underlying problems that are not merely a reflection of recent East-West tensions and mounting U.S. foreign trade deficits, let alone the product of one or two administrations." Among the underlying problems identified by the study panel were the following:

- The U.S. system of export administration lacks an effective means of weighing the risks of aiding the Soviets in enhancing their military and industrial capabilities against the potential costs of lost economic opportunities . . . economic costs of control versus the security risks of noncontrol are not well defined.

- In the absence of sound criteria of what is critical and feasible to control, the present control system is overreaching in scope, inconsistent in application, and overreliant on unilateral restrictions.
- The innovative capacity of the United States is regarded as a separate issue, unaffected by and having little bearing upon export control policy. In fact, accelerating the development and deployment of new weapon technologies would serve to protect U.S. security at the point where it is most vulnerable and to relieve some of the burden being placed on export controls as a means of preserving a diminishing technological advantage.[68]

The report acknowledged that the Soviet Union profited in various ways from the acquisition, duplication, and assimilation of certain Western technologies—such as computers and microelectronics—but cautioned that "broad generalizations about the contribution of technology transfer to Soviet weapon capabilities and production capacity are of limited value."[69] In general, it was thought that "national security export controls should be limited to goods and technologies that are critical to military capabilities likely to be crucial in preventing or fighting a war and whose acquisition could enable the Soviet Union to overcome deficiencies or surpass U.S. capabilities in the near term."[70] Significantly, the panel indicated that the development of near-term and critical criteria to identify technologies appropriate for export regulation would "appear to rule out controls on scientific communication and on emerging civilian technologies with conceivable but not immediate military application."[71]

The panel made no recommendations in its report that would require new legislative authority. It proposed that the President appoint a distinguished national study committee or commission to identify the militarily critical technologies in which the United States had the lead and that the nation needed to and actually could protect for reasons of maintaining its security. This national study panel would appoint certain expert task forces to develop ways to restrict Soviet access to the critical technologies identified by the panel for protection. The President and the pertinent committees of Congress would then pursue the implementation of the recommendations of these advisory bodies with the responsible departments and agencies. The report also expressed support for military procurement reforms initiated by Deputy Secretary of Defense Frank Carlucci, which evolved into the DOD Acquisition

Improvement Program. "Accelerating the deployment of new military equipment," said the report, "would serve to protect U.S. technological leadership at the point where it is most vulnerable."[72] However, the recommended advisory panels were not created, and the CSIS report received little media or official notice.

The next study report, *Export Controls: The Need to Balance National Objectives*, appeared in January 1986. It was prepared under the sponsorship of the Business-Higher Education Forum and was the fourth in a series addressing various aspects of American industrial competitiveness.[73] The series grew out of the Forum's premier assessment, *America's Competitive Challenge: The Need for a National Response*, delivered to President Reagan in May 1983.[74] Founded in 1978 in affiliation with the American Council on Education, the Business-Higher Education Forum "was created for the express purpose of promoting discourse and acting on issues shared jointly by American business and the nation's higher education institutions." Composed of "leading corporate and academic chief executives," the Forum provides "an opportunity for interchange among its members on matters pertaining to their respective interests and, hence, to the common aspirations and needs of the business and academic communities."[75]

In assessing the application and effect of export controls, the Forum's study foresaw a need to balance national objectives, to reconcile "the inevitable conflict between our societal interest in security, on the one hand, and our interests in technological innovation, economic progress and open communication, on the other."[76] The report acknowledged the existence of well-financed and organized East bloc efforts to acquire Western technology to gain military power and manufacturing advantages. However, although conceding the difficulty of making such judgments, the report also recognized "the Soviet Union has had trouble developing, disseminating and applying technology."[77] And even though Western technology may be more effectively used in Soviet military systems, said the report, "the Forum is aware of no case in which the transfer of U.S. technology has had significant strategic implications."[78]

The study found that export controls generally have an adverse effect on American high-technology industry by limiting sales, with consequential reductions of revenues and employment, and "by increasing the cost of carrying out those transactions that are permitted."[79] Moreover, the restraining effect of export controls on the flow of information was regarded as a particularly significant cost. "To the extent that controls to 'protect' technology also serve to isolate it and curtail its application," said the report, "the forces that propel innovation are weakened."[80] Similarly, export restrictions on scientific communication also were seen "to impede scientific advance," to prevent the development of new technology of commercial and military significance, "to drive the best scientists from the very fields that are determined to be of greatest national-security importance," and to be contrary "to the fundamental tenets of a free society."[81]

Among the recommendations offered in the Forum report were narrowing the militarily critical technologies subject to export control to a currently vital small set, maintaining timely and meaningful knowledge of the foreign availability of critical goods and technologies, minimizing controls on West-West trade, allowing "all communications concerning non-classified scientific research to be free from governmental restraints of any type," congressional limitation of the authority it delegated to the President to impose export controls for reasons of foreign policy, minimizing interagency review of individual export license applications, and authorizing judicial review of the administration of export controls.[82] Only a couple of these proposals received meaningful policy recognition: Executive branch capability for assessing foreign availability, for example, was subsequently enhanced, and judicial review of the administration of export controls was included in the Export Administration Act reauthorization vetoed by President Bush in November 1990.[83]

A special study panel of the National Academy of Sciences issued a report—*Balancing the National Interest*— in March 1987, following up on the 1982 Corson study.[84] Chartered through the Academy's Committee on Science, Engineering, and Public Policy, the Panel on the Impact of National Security Controls on International Technology Transfer was chaired by Lew Allen, Jr., the vice president of the California Institute of Technology, who was also the director of the Jet Propulsion Laboratory, former chief of staff of the Air Force and one-time director of the National Security Agency. Former members of the Corson panel serving on the new study committee were John M. Deutch, Mary L. Good, Franklin A. Lindsay, and Richard A. Meserve. Previously, Meserve had also assisted with the Business-Higher Education Forum study of export controls. The Allen group's support staff included Stephen A. Merrill, who had worked with the CSIS Panel on Technology Transfer, and were led by Mitchell B. Wallerstein, a former Corson committee staff consultant.[85]

Organized in 1984 "to examine the effect of export controls on commercial trade in high technology goods and information and on the vigor of U.S. high-technology industry,"[86] the Allen panel was tasked to "seek strategies to regulate the international transfer of technology through industrial channels in such a manner as to balance the national objectives of national security, economic vitality, scientific and technological advance, and commercial, educational, and personal freedom."[87] Academy leaders had been alerted to the commercial dimension of government restriction of scientific and technological information flow in 1982 by the Corson panel, which did not have the opportunity to examine this area in depth. Furthermore, the Academy was actively exploring ways in which the nation's capacity for technological innovation and, consequentially, its international trade position might be strengthened.[88] Finally, with the failure of the 98th Congress to reform and reauthorize the Export Administration Act, concern about the polarization of attitudes within government and industry over national security export controls was growing. The Allen panel was a response to all of these challenges—to examine the national security implications of technology transfer beyond the basic research stage, to seek a better balance between national interests in military security and beneficial overseas trade, and to open dialogue and promote consensus on these national interests among American industrial and governmental leaders.

In conducting its study, the Allen panel reviewed both open and officially restricted literature, invited testimony by representatives of pertinent agencies—the Departments of Defense, Commerce, State, Treasury, and Justice, and the National Aeronautics and Space Administration—and received three security classified briefings from the federal intelligence community.[89] Views were received from an array of industrial leaders and other individuals having expertise concerning various aspects of national security export controls. Fact-finding delegations of the panel visited six European countries—Austria, Belgium, France, Great Britain, Sweden, and West Germany—in January 1986. Other delegations traveled to five Asian nations—Hong Kong, Japan, Malaysia, Singapore, and South Korea—in March 1986. Finally, both outside consultants and the panel's professional staff prepared research papers that were published in the final report or a companion volume.[90] Support for the project was provided by federal agencies—the Departments of Commerce, Defense, Energy, and State, the National Aeronautics and Space Administration, and the National Science Foundation—over a dozen private organizations, and the National Research Council Fund.[91]

After reviewing intelligence information, the Allen panel agreed with "previous reports that the Soviet technology acquisition effort is massive, well financed, and frequently effective."[92] Furthermore, it was disclosed that, in the judgment of senior Western intelligence officials, the most valuable technology acquisition channel to the Soviets in enhancing their military capability was espionage, followed by diversion of unclassified but controlled technology, and then the transfer of uncontrolled Western technology. "The U.S. national security export control regime," observed the report, "does not provide solutions to the problem of espionage."[93]

Qualified acceptance of the Farewell evidence was expressed. It was noted that such internal Soviet documents "must be viewed with caution (because of the possibility that the authors had an interest in inflating the successes of the acquisition program in their reports to superiors)." Nonetheless, the report said "the Farewell papers do set out a remarkable record of the scope and success of the Soviet acquisition effort" and called these materials "solid evidence on Soviet technology acquisition."[94]

The Allen panel and "other experts it consulted" found the conclusions and the underlying research methods of a highly touted February 1985 Pentagon study—*Assessing the Effect of Technology Transfer on U.S./Western Security: A Defense Perspective*—"unconvincing."[95] A unique attempt to quantify the benefits of export controls, the assessment, utilizing rejected 1983–1984 export license applications, "estimated that the Soviets could have saved $0.5 billion to $1 billion a year over a 13-year period if the applications in the sample had been approved and the acquired technology exploited."[96] Finding DOD cost savings assertions unsupported in the text of the report and lacking requested backup data for these figures, the Allen panel questioned "how much weight these estimates should be accorded."[97]

The panel noted the stultifying effects that secrecy had on Soviet scientific and technological development, and, although observing that the Soviets had experienced difficulty incorporating new scientific ideas into development and production in their civilian sector, acknowledged the great resource advantages that the military sector had to utilize Western scientific and technological knowledge effectively.[98]

Probing the competitive effects of export controls, the panel conducted an industry survey oriented toward aircraft, electronics, instrumentation, and machine tool firms and based on experience during the 12 months prior to May 1986. "The 170 respondents accounted for roughly $36 billion of foreign sales in 1985," said the panel report, "or approximately 28 percent of estimated total U.S. high-technology sales." Concerning adverse impact:

- 52% experienced lost sales due to export controls.
- 26% had Free World customers turn down business deals because of export controls.
- 38% reported existing customers indicated a preference to shift to sources of supply that would not entangle them in U.S. export controls.

- 50+% expected such occurrences to increase in number over the next 2 years.[99]

The panel also developed concise check lists of the disadvantages to U.S. exporters vis-à-vis both COCOM and non-COCOM competitors.[100]

Reminiscent of Harold Lasswell's integrated national security policy-making procedure, discussed earlier, was the panel's position that "U.S. policy on national security export controls should result from a process that weighs the benefits of controls to the United States in its relations with adversaries against the costs of controls in relations with allies and trading partners."[101] The existing arrangement was found to be deficient, lacking no regular policy guidance from the highest government levels and no effective means for reconciling agency conflicts, in want of resources for and assertiveness by key institutions, such as the Departments of Commerce and State, and suffering from a concentration of authority in narrow interest administrative units with little or no technology and trade development perspective. However, said the report, "the panel believes that both the policy guidance and the division of labor set forth in the Export Administration Act are appropriate."[102]

Some of the other findings and key judgments of the Allen panel were that pragmatic export control lists had to be technically sound, narrowly focused, and multilaterally coordinated; controls on the employment of foreign nationals in the American research and development community had to be selectively and sparingly used; current export licensing requirements, security classification procedures, and proprietary controls for technical data were both appropriate and adequate; integration of the Militarily Critical Technologies List into the Commodity Control List could not be accomplished; the complexity of U.S. export policies and procedures discouraged compliance; the federal government did not maintain adequate information to evaluate the impact of national security export controls; and a comprehensive cost-benefit analysis of export controls was currently infeasible.[103]

The panel made a number of recommendations, all within the context of what were thought to be appropriate national policy objectives. These, said the panel, were promotion of "the economic vitality of Free World countries," maintenance and invigoration of the domestic technological base, and cooperation with allies to impede East bloc acquisition of Western technology enhancing military capability. Several specific recommendations were offered under the goals of strengthening the COCOM mechanism and according greater importance to the maintenance of U.S. technological strength, economic vitality, and allied unity when making national security export control decisions. With regard to the latter goal, the panel recommended allocating sufficient resources and authority for the Departments of Commerce and State to fulfill their roles in export operations, bringing overall balance to the national security export licensing process with greater emphasis on technical judgment, rescinding the statutory requirement to integrate the MCTL into the Commodity Control List, and establishing a mechanism for realizing effective, two-way communication between the highest levels of government and the private sector concerning the formulation and implementation of U.S. export policy. The panel also urged Congress to give the Department of Commerce both the resources and the obligation to develop reliable information on the operation and impact of national security export controls, and urged, as well, that pertinent intelligence be collected and better used, including a public education dimension.[104]

The Allen panel's report received initial congressional scrutiny in early February 1987 at a hearing before the House Subcommittee on International Economic Policy and Trade. At the time, efforts were underway to develop legislation that would eventually become the Omnibus Trade and Competitiveness Act of 1988.[105] Appearing before the Subcommittee to discuss the report was panel chairman Lew Allen, Jr., accompanied by Tom Christiansen, Richard Meserve, Leif Olsen, and project director Mitchel Wallerstein. Various aspects of the report were discussed and, generally, it was favorably regarded.[106]

The following day, the Allen panel report was discussed at a hearing before the House Committee on Science, Space, and Technology. Introductory remarks were offered by Frank Press, president of the National Academy of Sciences, and Robert White, president of the National Academy of Engineering. Appearing on behalf of the panel were Lew Allen, Jr., Tom Christiansen, Leif Olsen, Ruth Greenstein, and Mitchel Wallerstein. Other than a minor concern that the panel might have devoted more attention to the leakage of American technology through South America, the Allen panel's report was well received.[107] Due to time constraints, Assistant Secretary of Defense Richard Perle, originally scheduled to testify at the same hearing, could not be heard for several weeks. In late April, Perle returned. In the opening paragraphs of his prepared statement, he commended the Allen panel "for addressing the question of increasing U.S. exports," but found its study "did not have the perspective to judge the relative importance of export controls and other factors in our economic performance." He saw export controls as, "at worst, only one factor in the trade deficit" and took issue with panel's claim that Pentagon review of one export license application took over 900 days to be decided.

Nonetheless, he agreed "with a number of recommendations made by the panel." Specifically cited were proposals that the United States take the lead in further strengthening the COCOM system and "the idea of negotiating comprehensive understandings with third countries."

He also indicated "disappointment at several recommendations of the National Academy study" and disagreement with some of its conclusions—for example, "that the Defense Department has taken an overly active role in export controls." Finally, in conclusion, Perle lamented "that the Academy report does not even discuss the substantial accomplishments of the Administration in stopping the loss of technology to the Soviet bloc."[108] In the rather rambling discussion that ensued, he conceded that, in terms of his general criticism, the Allen panel "probably construed their mandate a little more narrowly than I would have, if the point was to look at the overall picture,"[109] and that the report "does not put sufficient emphasis on the extent to which the problem that it discusses could be resolved by more effective administration of the existing system." According to Perle:

> It places too much emphasis on removing commodities and technologies from control, which I think has to be done but on a discriminating basis and not on a wholesale basis; and I think it underestimates the extent to which we could improve the competitiveness of American industry by the timely delivery of licenses that had been subjected to a serious review on security grounds, but because that review has been conducted by organizations that are modern and competent, it can be done in a reasonable period of time and indeed it can be done as quickly as it is done in some other countries.[110]

Subsequently, in February 1989, the staff of the House Committee on Science, Space, and Technology prepared a report summarizing and analyzing the hearings on the Allen panel's findings and recommendations. Indicating it was "not prepared to endorse any specific recommendations in total at this time,"[111] the committee took issue with the Allen panel's quantitative analysis of the impact of U.S. national security export controls on American competitiveness.[112] Indeed, "the Committee staff developed its own assessment, resulting in job loss estimates only 15% to 20% of the 188,000 presented in the Academy report."[113] Nonetheless, others viewed the Allen panel's costing efforts as an original and valuable contribution, and an important counterbalance to Pentagon calculations of huge military costs resulting from not having export controls.[114]

Another pertinent Academy report appeared the following year. *Global Trends in Computer Technology and Their Impact on Export Control* was produced by the Committee to Study International Developments in Computer Science and Technology. The panel, which operated during late 1987 and through the summer of 1988, was mandated by the Computer Science and Technology Board of the National Research Council to assess computer technology development and international diffusion trends and their ramifications for export control policy. Members of the study panel were selected on the basis of their expertise concerning a variety of aspects of computer science and technology, international trends in the field, and knowledge of pertinent East bloc conditions.[115] The panel was chaired by Seymour Goodman, professor of management information systems and policy at the University of Arizona's College of Business and Public Administration.

The study panel's report consisted of "a set of individual technology assessments focusing on conditions in CoCom countries," computing in East bloc nations, and certain conclusions and recommendations. Following an introductory overview, subsequent chapters examined computer hardware, the manufacturing of computer systems, software, computer networks, and the particulars of computing in East bloc countries. The panel concluded that:

- Export controls were made more difficult by prevailing technological trends, including the generally rapid development of computer technology.
- Manufacturing and development know-how were "fundamental" to the computer technology advantage of COCOM nations.
- Commercial vitality is "essential" to technical vitality.
- Nations outside of COCOM "are also expanding the supply of computer technology."
- Although East bloc prospects in computer technology capabilities were improving, they were, nonetheless, still "constrained."[116]

Among the recommendations offered, the report urged better definition of controlled computer technologies, clarification of and elimination of controls on inter-COCOM trade in certain computer technologies, effective elimination of controls on computer technologies available from sources outside of COCOM, formulation of a U.S. policy "for preventing computer networks from becoming a channel for significant covert technology transfer and to protect the computational resources of CoCom countries," and focusing federal export controls on "computer technologies of compelling military importance" that could enable East bloc countries "to make substantial gains . . . ."[117] There appeared to be, however, little congressional or public notice of the panel's report.

When legislating the Omnibus Trade and Competitiveness Act of 1988, Congress mandated still another evaluation, once again to be performed under the auspices of the National Academy of Sciences complex. The statute mandated the Secretaries of Commerce and Defense to arrange with the National Academy of Sciences for the conduct of "a comprehensive study of the adequacy of the current export administration system in safeguarding United States national security while maintaining United States international competitiveness and Western technological preeminence." Specifically, the study was to:

- Identify those goods and technologies which are likely to make crucial differences in the military capabilities of controlled countries, and identify which of those goods and technologies controlled countries already possess or are available to controlled countries from other sources.
- Develop implementable criteria by which to define those goods and technologies.
- Demonstrate how such criteria would be applied to the control list by the relevant agencies to revise the list, eliminate ineffective controls, and strengthen controls.
- Develop proposals to improve United States and multilateral assessments of foreign availability of goods and technology subject to export controls.
- Develop proposals to improve the administration of the export control program, including procedures to ensure timely, predictable, and effective decision-making.[118]

The Academy's Committee on Science, Engineering, and Public Policy (COSEPUP) undertook the formulation of the study panel, which was formally organized in August 1989.[119] The Panel on the Future Design and Implementation of U.S. National Security Export Controls began its work in the fall of the year under the leadership of Roland W. Schmitt, president of Rensselaer Polytechnic Institute and the recent chairman of the National Science Board. Among its members were William Perry from the Corson study panel, Herbert Dwight, Jr., Ruth Greenstein, and John McLucas from the Allen study panel, and Seymour Goodman, who had chaired the National Research Council's *Global Trends* panel on the international availability of computer technology.[120] Furthermore, directing the new study panel's staff was Mitchel Wallerstein, the former staff director of the Allen panel, who had also worked with the Corson panel.

Like its Academy predecessors examining national security controls on scientific communication and technology transfer, the Schmitt panel had presentations from representatives from the Departments of Defense, Commerce, State, Treasury, and Energy, the National Aeronautics and Space Administration, and the Office of Science and Technology Policy. The full panel was also given security classified briefings by the Technology Transfer Intelligence Committee, an interagency unit of the federal intelligence community, and some working groups of the panel selectively received additional government briefings at high levels of classification. Some panel members participated in fact-finding missions. A group visited Hong Kong, Japan, Macao, South Korea, and Taiwan in February 1990. Another delegation went to Belgium, France, Germany, Switzerland, and the United Kingdom in May 1990. A third group visited Canada. Expert subpanels on advanced industrial materials, commercial aircraft and jet engines, and computer technology, led by members of the Schmitt panel and composed of other knowledgeable individuals, offered their special findings, as did research prepared by staff and outside consults, for the study panel's consideration.[121]

Although the panel formally ended its study in late 1990, changing world conditions—particularly in Eastern Europe and the Soviet Union—continued to challenge panel members and staff attempting to prepare a final report. It finally appeared in late January 1991 with little public notice and little interest on the part of Congress, which had been struggling for months with amendments revising and reauthorizing the Export Administration Act.[122]

The final report offered various key findings and conclusions, not the least of which were the following.

- Export controls, somewhat modified, were a "necessary and appropriate policy instrument" for responding to national security threats, and should be sufficiently flexible to address "new types and sources of national security challenges, such as those derived from growing proliferation threats or the threat of terrorism, as they emerge."[123]
- There is a "continuing paucity of reliable data" on both Soviet efforts to acquire U.S. technology since 1989 and the acquisition efforts of "countries that are the focus of proliferation concern."[124]
- Various factors—U.S. defense industrial base problems, shifts from defense to commercially driven innovation, new Asian and European industrial competitors, and the growing importance of export controls for the U.S. economy—have resulted in greater awareness of the importance of economic factors in the development of U.S. national security policy.[125]
- Among the basic problems in U.S. export control policy are a multiplicity of diverse and legislatively and administratively

uncoordinated statutes, agencies, and regimes, resulting in jurisdictional disputes, overlapping enforcement, outdated and confusing control lists, ineffective dispute resolution, counterproductive unilateral export controls, insufficient judicial review, and inadequate industry participation.[126]

- Substantial reform of the U.S. export control regime "will be necessary to achieve the goals of an effective U.S. export control process."[127]

A number of recommendations were offered by the panel, including the following proposals:

- Changing international conditions affecting U.S. national security—particularly in Eastern Europe and the Soviet Union—warranted modification of export control policy regarding those nations to "reflect a mixed strategy of encouraging political and economic reform and making further relaxation of controls contingent on evidence of additional change."[128]
- There should be an expansion of federal intelligence community assessment of current Soviet technology acquisition and utilization efforts and wider distribution of these assessments to relevant federal agencies and to COCOM allies.[129]
- There should be reconsideration of the continued need for dual use export controls as conditions change in the Soviet Union and/or other currently proscribed countries.[130]
- Among the reforms that should be undertaken by the executive branch through the issuance of a presidential national security directive, available to the public in as much detail as possible, were (a) "guidance on the fundamental objectives for all national security export controls"; (b) "general policy and specific procedures for constructing and reviewing the control lists"; (c) guidance for the development of regulatory control regimes"; (d) creation of formal policy mechanisms, such as an Export Control Policy Coordinating Committee, interagency working groups, and technical groups; (e) inclusion of federal economic agencies in the national security policy process "when their expertise could serve the national interest"; (f) routine inclusion of the Secretary of Commerce as an advisor/participant in National Security Council discussions; (g) development of standards for munitions and dual use items; and (h) "guidance on timely procedures for resolving interagency disputes."[131]
- Other steps to be taken for better policy execution included (a) "consolidation of all day-to-day administrative functions in a

single agency"; (b) designating, with appropriate upgrading of resources, the Bureau of Export Administration of the Commerce Department as the single administrative agency for export controls; (c) fashioning a set of integrated U.S. control lists; (d) enhancing U.S. representation at COCOM; and (e) increasing industry participation in the export control system.[132]

COMPETITIVENESS

Although its report did not command the attention or exert the immediate influence that the Schmitt panel undoubtedly desired, its assessment of changing U.S. national security export control interests in a period of major power modification in the world was pivotally perceptive. Calling the U.S. national security export regime, as well as the COCOM multilateral control framework, an "artifact" of the "now ended" Cold War, the panel foresaw "the most difficult and urgent challenges" to American national security as being economic and technological rather than "purely military in nature." This assessment did not ignore the remaining threat posed by "the size and configuration of the Soviet Union's strategic nuclear arsenal and its continuing modernization, together with still sizable Soviet conventional ground forces," or "a generally heightened potential for regional hostilities," such as the recent intervention by the United States and its allies in the Persian Gulf.[133] Other developments, however, were seen as being more critical for U.S. national security interests, developments for which current export controls may be of limited utility in achieving national security objectives.[134]

The Schmitt panel identified the following as being among the most significant economic and technological challenges facing the United States in the aftermath of the Cold War:

- *Changing global economy structure.* Revolutionary developments in information and telecommunications technologies have afforded multinational corporations, operating in worldwide markets, great flexibility in deciding where to locate research and development and manufacturing facilities, with resulting U.S. loss of capacity in some key technology areas and a blurring of the national identities of such firms and their technologies.
- *Accelerated global diffusion of technology.* As multinational companies constantly explore new international markets and site their facilities in various countries, they contribute to the increasingly rapid diffusion of technology. So, too, does the growing tech-

nology development and production capacity of many geographically dispersed nations.

- *Declining U.S. preeminence.* The technological and manufacturing preeminence enjoyed by the United States since World War II has receded in recent years and continues to subside. Furthermore, as a consequence of aging, segments of the nation's research and development and manufacturing infrastructure require significant infusions of new capital.
- *Growing East Asian competition.* Rapidly developing technological and manufacturing sophistication in Japan, as well as Hong Kong, Singapore, South Korea, and Taiwan, has fueled their seizure of major shares of the U.S. market for steel, automobiles, semiconductors, and consumer electronics. Japan's rapid rise to the position of an economic superpower has been assisted by a tax system and import control regime that encourage long-term growth over short-term profits.
- *Changing global economic and financial power distribution.* The world's leading debtor nation in 1986, the United States continues to suffer large negative trade balances with several of its major trading partners, including Japan, South Korea, and Taiwan. Japan's growing economic and financial power, by contrast, is evidenced by seven of the world's 10 largest banks now being Japanese.
- *U.S. defense industrial base weakening.* Small-to-medium-sized U.S. companies, due to, among other factors, low profits, disincentives in the federal procurement system, and other burdensome government regulations, are abandoning the defense sales sector in large numbers. Pentagon dependence on foreign sources for goods and technologies has various national security implications, including supply vulnerabilities and contributing to the erosion of U.S. research and development and manufacturing capabilities.
- *Exports for U.S. economic vitality.* The nation continues to be the world's largest international trader and is becoming nearly as dependent on exports as are its major competitors.[135]

The response of the Schmitt panel to these new challenges generally was twofold: Give increased weight to economic factors in the formulation of U.S. national security policy *and* overhaul the national security export regime along the lines recommended, with minimal impediment of U.S. international trade and export controls being set, through multinational cooperation, on items with particular end uses that pose immediate or near-term threats. Although East-West relations were dramatically different from what they had been 10, five, or only two years ago, both of these points reiterated still valid positions expressed from within the scientific community in opposition to increased national security controls on traditional scientific communication during the 1980s. In the first instance, national security, in its most meaningful context, should not be defined by the military capability of hostile nations alone, but should be broadly understood. Furthermore, regarding the Schmitt panel's second point of response, national security is realized by achievement in science, not by concealment. National security results from the broad distribution of scientific and technological knowledge, both for purposes of advancing such knowledge through criticism and improvement and for economic gain through sales in open markets at home and abroad.

Scientific and technological knowledge, however, can have societal repercussions and, therefore, where risks to a nation's security may be involved, limited exceptions to these premises should be expected. Once again, we are reminded of Lasswell's call for a procedure to create such security policy.[136] The Constitution of the United States provides one such procedure—for creating statutory law. The Schmitt panel recommended an additional procedure—for realizing better coordination and administration of legislated national security export controls.[137]

These, however, are but one type of authority among many that have been invoked and applied of late to restrain the flow of scientific and technological knowledge. Whereas the oldest of them trace their origins to the World War I era, all owed their continued existence to the prevalence of the Cold War. Yet, that era is over because the international conditions creating and sustaining it have ended. What significance does the legacy of national security controls have in the new world order?

NOTES AND REFERENCES

[1]Thomas P. Hughes. *American Genesis.* New York: Viking Penguin, 1989, p. 170.

[2]See, generally, Thomas W. Langfitt, Sheldon Hackney, Alfred P. Fishman, and Albert V. Glowasky, eds. *Partners In The Research Enterprise.* Philadelphia: University of Pennsylvania Press, 1983; U.S. Congress. House. Committee on Science and Technology. *Commercialization of Academic Biomedical Research.* Hearings, 97th Congress, 1st Session. Washington, DC: U.S. Govt. Print. Off., 1981; U.S. Congress. House. Committee on Science and Technology. *Joint Industry/University Cooperation With Federally Supported Research Facilities.* Hearing, 98th Congress, 1st Session. Washington, DC: U.S. Govt. Print. Off., 1983; U.S. National Science Foundation. *University-Industry Research Relationships: Myths, Realities, and Potentials.* Fourteenth Annual Report of the National Science Board to the President and the Congress. Washington, DC: National Science Foundation, 1982;

U.S. National Science Foundation. National Science Board. *University-Industry Research Relationships: Selected Studies.* Washington, DC: National Science Foundation, 1983; Vivian Weil and John W. Snapper, eds. *Owning Scientific and Technical Information: Value and Ethical Issues.* New Brunswick, NJ: Rutgers University Press, 1989.

[3]Hughes, *American Genesis,* p. 15; on the earlier period, see David Freeman Hawke. *Nuts and Bolts of the Past.* New York: Harper and Row, 1988.

[4]*Ibid.*, p. 7.

[5]See *Ibid.*, pp. 24, 53, 70.

[6]*Ibid.*, pp. 48–49.

[7]*Ibid.*, pp. 7–8, 53–54, 96, 138–139, 183. Industry's greater ability and willingness to challenge the patents of independent inventors and protect its own during the years after World War I also contributed to the demise of the inventors. *Ibid.*, p. 139.

[8]*Ibid.*, pp. 53, 138–139, 160, 172, 175, 180.

[9]U.S. Department of Defense. Office of the Director of Defense Research and Engineering. *An Analysis of Export Control of U.S. Technology—A DOD Perspective: A Report of the Defense Science Board Task Force on Export of U.S. Technology.* February 4, 1976. Washington, DC, p. 1; for background on the Defense Science Board, see Chapter 4, note 55.

[10]U.S. Department of Defense. Office of the Director of Defense Research and Engineering. *An Analysis of Export Control of U.S. Technology—A DOD Perspective,* p. 3.

[11]93 Stat. 503.

[12]U.S. Department of Defense. Office of the Director of Defense Research and Engineering. *An Analysis of Export Control of U.S. Technology—A DOD Perspective,* pp. 5–6; also see J. Fred Bucy. "Technology Transfer and East-West Trade: A Reappraisal." *International Security,* v. 5, Winter, 1980–1981, p. 135.

[13]Paul Wallich, ed. *Managing The Flow of Technical Information: An Industry/Government Dialogue.* Sponsored by *IEEE Spectrum* Magazine. June 2, 1982. Washington, DC, p. 10.

[14]Stuart Macdonald. *Technology and the Tyranny of Export Controls.* Houndmills: The Macmillan Press, 1990, p. 60.

[15]U.S. Congress. Senate. Committee on the Judiciary. *Information and Competitiveness.* Hearing, 100th Congress, 2d Session. Washington, DC: U.S. Govt. Print. Off., 1989, pp. 9–10, 13.

[16]For the table of contents of the Militarily Critical Technologies List, as initially compiled, see *Federal Register,* v. 45, October 1, 1980, pp. 65014–65019; the Department of Energy, in support of the Department of Defense, prepared a list of energy-related militarily critical technologies, but only its table of contents was published in *Ibid.*, pp. 65152–65167; an unclassified version of the MCTL was produced in October 1986.

[17]Richard Barnard. "Pentagon Mulls New Technology Secrets List." *Defense Week,* v. 4, October 17, 1983, pp. 1, 5; Richard Barnard. "Pentagon Weighs New Strictures On Unclassified Papers." *Defense Week,* v. 4, October 24, 1983, pp. 1, 10-11.

[18]As quoted in Willie Schatz. "The Hitch in High-Tech Trade." *Datamation,* v. 29, October, 1983, pp. 148–159.

[19]"Technology Transfer: A Policy Nightmare." *Business Week,* April 4, 1983, p. 95.

[20]Paul Wallich, ed. *Managing The Flow of Technical Information: An Industry/Government Dialogue,* p. 10.

[21]U.S. Department of Defense. Office of the Director of Defense Research and Engineering. *An Analysis of Export Control of U.S. Technology—A DOD Perspective,* p. 4. At this point in the report, it was stated: "The Task Force believes that these factors exist in Russia and Eastern Europe, making them receptive hosts for any active efforts to transfer those technologies studied by the subcommittees." Technologies studied by the Task Force's subcommittees included airframes, aircraft jet engines, instrumentation, and solid state devices.

[22]Bucy, "Technology Transfer and East-West Trade: A Reappraisal," p. 136.

[23]Macdonald, *Technology and the Tyranny of Export Controls*, pp. 167, 171–172, 193.

[24]William L. Long. *U.S. Export Control Policy*. New York: Columbia University Press, 1989, p. 23.

[25]*Ibid.*, p. 24.

[26]*Ibid.*, p. 31.

[27]83 Stat. 841.

[28]Long, *U.S. Export Control Policy*, p. 33.

[29]Macdonald, *Technology and the Tyranny of Export Controls*, p. 35.

[30]I.M. Destler. "The Evolution of Reagan Foreign Policy" in Fred I. Greenstein, ed. *The Reagan Presidency: An Early Assessment*. Baltimore: The Johns Hopkins University Press, 1983, p. 119.

[31]*Ibid.*, pp. 119–120.

[32]*Ibid.*, p. 121.

[33]*Ibid.*, p. 122.

[34]*Ibid.*, p. 154.

[35]Samuel P. Huntington. "The Defense Policy of the Reagan Administration, 1981–1982" in Fred I. Greenstein, ed. *The Reagan Presidency: An Early Assessment*, p. 93.

[36]*Ibid.*, p. 94.

[37]Chicago Tribune. "Military Eyed to Police High-Tech Shipping." *Washington Post*, September 26, 1982, p. A15.

[38]Stuart Auerbach. "Pentagon's Export Role Is Expanded." *Washington Post*, March 22, 1984, pp. B1, B6; Clyde H. Farnsworth. "Pentagon's Wider Role On Exports." *New York Times*, March 21, 1984, pp. D1, D23; Eduardo Lachica. "New U.S. Technology Export Guidelines Broaden Pentagon's License Review Role." *Wall Street Journal*, March 22, 1984, p. 10; Michael Schrage. "Pentagon Wins Trade Fight." *Washington Post*, March 24, 1984, pp. A1, A6.

[39]See, for example, the testimony offered in U.S. Congress. Senate. Committee on Baking, Housing, and Urban Affairs. *East-West Trade and Technology Transfer*. Hearing, 97th Congress, 2d Session. Washington, DC: U.S. Govt. Print. Off., 1982.

[40]For popular accounts of such activity, see Linda Melvern, David Hebditch, and Nick Anning. *Techno-Bandits*. Boston: Houghton Mifflin Company, 1984; Robyn Shotwell Metcalfe. *The New Wizard War*. Redmond, WA: Tempus Books, 1988; Lauran Paine. *Silicon Spies*. New York: St. Martin's Press, 1986.

[41]See Associated Press. "Leakage of U.S. Technology To the Soviet Bloc Is Charged." *New York Times*, May 6, 1982, p. A10; Paul Mindus. "Turf Squabbles and Inexperience Hamper Technology Guardians." *Washington Post*, October 27, 1983, p. A21; Edward T. Pound. "Curbs on Technology Exports Hurt by Gaps in Enforcement." *New York Times*, October 14, 1981, pp. A1, A24; U.S. Congress. Senate. Committee on Governmental Affairs. *Transfer of United States High Technology to the Soviet Union and Soviet Bloc Nations*. S. Rept. 97–664, 97th Congress, 2d Session. Washington, DC: U.S. Govt. Print. Off., 1982.

[42]U.S. General Accounting Office. *Export Control Regulation Could Be Reduced Without Affecting National Security*. GAO/ID-82-14. Washington, DC: U.S. Govt. Print. Off., 1982, p. ii.

[43]See Macdonald, *Technology and the Tyranny of Export Controls*, pp. 35–39.

[44]*Ibid.*, pp. 33–34.

[45]See Michael Weisskopf. "Report Shows How Soviets Amass Western Technology." *Washington Post*, September 19, 1985, pp. A1, A15.

[46]Director of Central Intelligence. *Soviet Acquisition of Militarily Significant Western Technology: An Update*. September, 1985. Washington, DC, preface. A more general report on the same subject matter, *Soviet Acquisition of Western Technology*, had been issued in April 1982.

[47]Director of Central Intelligence. *Soviet Acquisition of Militarily Significant Western Technology: An Update*, p. 2.

[48]*Ibid.*, pp. 17–24.

[49]*Ibid.*, p. 2.

[50]*Ibid.*, pp. 1, 24.

[51]Weisskopf, "Report Shows How Soviets Amass Western Technology;" National Academy of Sciences. *Balancing The National Interest*. Washington, DC: National Academy Press, 1987, pp. 41–42.

[52]Director of Central Intelligence. *Soviet Acquisition of Militarily Significant Western Technology: An Update*, p. 17. Curiously, in early 1985, Secretary of Commerce Malcolm Baldridge complained in a letter to the President's national security adviser, the Secretaries of State, Defense, and Energy, and the Administrator of the National Aeronautics and Space Administration that the failure of their staffs to check the sensitivity of documents before their submission to the National Technical Information Service "permits the Soviets to acquire tens of thousands of scientific and technical studies as well as other strategic information." Baldridge, according to his letter, had been seeking better cooperation on this matter since 1982. Stuart Auerbach. "Baldridge Warns of 'Giveaway' Of Strategic Secrets to Soviets." *Washington Post*, February 21, 1985, pp. A1, A12; Martin Tolchin. "U.S. Said To Allow Secrets Giveaway." *New York Times*, February 24, 1985, p. 4.

[53]Michael Schrage. "U.S. May Tighten Electronic Net to Control Software." *Washington Post*, May 6, 1984, pp. D1, D3.

[54]John Horgan. "Air Force Proposal to Restrict Databases Faces Stiff Opposition." *The Institute*, v. 10, August, 1986, pp. 1–2.

[55]For background on such national security policy instruments, see Harold C. Relyea. "The Coming of Secret Law." *Government Information Quarterly*, v. 5, no. 2, pp. 97–116.

[56]An unclassified version of NSDD 145 is reproduced here in the appendix of documents.

[57]Diane Fountain. Transcript of remarks. National Information Policy Forum sponsored by Information Industry Association. November 11, 1986. New York. Substantially the same quotation may be found in "Are Data Bases A Threat To National Security?" *Business Week*, December 1, 1986, p. 39.

[58]NTISSP No. 2, "National Policy on Protection of Sensitive, But Unclassified Information in Federal Government Telecommunications and Automated Information Systems." October 29, 1986. Section 2. The text of this directive is reproduced here in the appendix of documents.

[59]See "Are Data Bases A Threat To National Security." *Business Week*, December 1, 1986, p. 39; Associated Press. "Pentagon Weighs Data Bank Curbs." *New York Times*, November 13, 1986, p. B11; Jerry J. Berman. "National Security Vs. Access To Computer Databases: A New Threat To Freedom of Information." *Software Law Journal*, v. 2, Winter, 1987, pp. 1–15; Ross Gelbspan. "Reagan Seeks Controls on Database Access." *Boston Globe*, April 20, 1987, pp. S5, S7; Michael Schrage. "U.S. Limits Access to Information Related to National Security." *Washington Post*, November 13, 1986, pp. A1, A29.

[60]See U.S. Congress. House. Committee on Government Operations. *Computer Security Act of 1987*. Hearings, 100th Congress, 1st Session. Washington, DC: U.S. Govt. Print. Off., 1987. Poindexter's involvement in the Iran-Contra affair is discussed in U.S. Congress. House. Select Committee to Investigate Covert Arms Transactions with Iran. [and] Senate. Select Committee on Secret Military Assistance to Iran and the Nicaraguan Opposition. *Report of the Congressional Committees Investigating the Iran-Contra Affair*. H. Rept. 100–433 and S. Rept. 100–216, 100th Congress, 1st Session. Washington, DC: U.S. Govt. Print. Off., 1987.

[61]101 Stat. 1724.

[62]See U.S. Congress. House. Committee on Government Operations. *Military and Civilian Control of Computer Security Issues*. Hearing, 101st Congress, 1st Session. Washing-

ton, DC: U.S. Govt. Print. Off., 1989; U.S. Congress. House. Committee on Science, Space, and Technology. *Implementation of the Computer Security Act.* Hearing, 101st Congress, 1st Session. Washington, DC: U.S. Govt. Print. Off., 1989; U.S. Congress. House. Committee on Science, Space, and Technology. *Implementation of the Computer Security Act (Public Law 100-235).* Hearing, 101st Congress, 2d Session. Washington, DC: U.S. Govt. Print. Off., 1990.

[63]See National Academy of Sciences. *Computers At Risk.* Washington, DC: National Academy Press, 1991.

[64]Paul Wallich. "Technology Transfer At Issue: The Industry Viewpoint." *IEEE Spectrum,* v. 19, May, 1982, p. 71.

[65]As identified in its report, the members of the CSIS Panel on Technology Transfer were as follows:

David Shore, Chairman –Division Vice President, RCA Corporation; member, CSIS Science and Technology Committee

Harold M. Agnew–President, GA Technologies; member, CSIS Science and Technology Committee

Jack Baranson–Center for Research on Industrial Strategy and Policy, Illinois Institute of Technology

J. Fred Bucy–President and Chief Executive Officer, Texas Instruments; member, CSIS Science and Technology Committee

John S. Foster, Jr.–Vice President, TRW; member, CSIS Science and Technology Committee

Benjamin Huberman–Vice President, The Consultants International Group; member, CSIS Science and Technology Committee

Donald Kerr–Director, Los Alamos National Laboratory

James A. McDivitt–Executive Vice President, Rockwell International; member CSIS Science and Technology Committee

John E. Schmick–Vice President, Boeing International

Frederick Seitz–President emeritus, Rockefeller University; Chairman, CSIS Science and Technology Committee

Edward Teller–Hoover Institution; member, CSIS Science and Technology Committee

Gus Weiss–consultant

John J. Welch–Senior Vice President, LTV Aerospace and Defense Company; member, CSIS Science and Technology Committee

[66]CSIS Science and Technology Committee. Panel on Technology Transfer. *Securing Technological Advantage: Balancing Export Controls and Innovation.* Stephen A. Merrill, ed. Washington, DC: Center for Strategic and International Studies, 1985, p. 8.

[67]*Ibid.*, pp. 8–9.

[68] *Ibid.*, p. 11.

[69]*Ibid.*, p. 16.

[70]*Ibid.*, p. 27.

[71]*Ibid.*, p. 28.

[72]*Ibid.*, pp. 40–43.

[73]Overseeing the export control project was Forum member Robert M. Rosenzweig, president of the Association of American Universities, with the assistance of attorneys Richard A. Meserve and Bruce D. Lindsay, both of Covington and Burling. Business-Higher Education Forum. *Export Controls: The Need to Balance National Objectives.* Washington, DC: Business-Higher Education Forum, 1986, p. iv.

[74]Other derivative study reports were *Corporate and Campus Cooperation: An Action Agenda,* published in May 1984; *The New Manufacturing: America's Race to Automate,* produced in June 1984; and *America's Business Schools: Priorities for Change,* appearing in May 1985. *Ibid.*, p. 56.

[75]*Ibid.*, p. 57.
[76]*Ibid.*, p. 1.
[77]*Ibid.*, p. 9.
[78]*Ibid.*, pp. 10–11.
[79]*Ibid.*, p. 13.
[80]*Ibid.*, p. 15.
[81]*Ibid.*, p. 34.
[82]*Ibid.*, pp. 4–7.
[83]See *Weekly Compilation of Presidential Documents*, v. 26, November 26, 1990, pp. 1839–1840.
[84]A draft of the panel's report, however, was leaked to the press and prematurely revealed in January 1987. See Stuart Auerbach. "Panel Hits U.S. Curbs on Exports." *Washington Post*, January 12, 1987, pp. A1, A16.
[85]As identified in the panel's final report, the members were as follows:

Lew Allen, Jr., Chairman–Vice President, California Institute of Technology, and Director, Jet Propulsion Laboratory; former Chief of Staff, U.S. Air Force; former Director, National Security Agency

Daniel Berg–President, Rensselaer Polytechnic Institute

Thomas A. Christiansen–Manager, International Trade Relations, Hewlett-Packard Company; member, Subcommittee on Export Administration, President's Export Council

W. Dale Compton–Senior Fellow, National Academy of Engineering; former Vice President of Research, Ford Motor Company

Richard N. Cooper–Boas Professor of International Economics, Harvard University; former Under Secretary of State for Economic Affairs

John M. Deutch–Provost, Massachusetts Institute of Technology; former Under Secretary of Energy

Herbert M. Dwight, Jr.–Chairman and Chief Executive Officer, Spectra-Physics, Inc.

Alexander H. Flax–President emeritus, Institute for Defense Analysis; former Assistant Secretary of the Air Force for Research and Development; former Chief Scientist, U.S. Air Force

John S. Foster, Jr.–Vice President, Science and Technology, TRW Incorporated; former Director, Defense Research and Engineering, Department of Defense; former Director, Lawrence Livermore National Laboratory; former Associate Director, Lawrence Berkeley National Laboratory

Frederick W. Garry–Vice President, Corporate Engineering and Manufacturing, General Electric Company

Mary L. Good–President and Director of Research, UOP Incorporated, Signal Research Incorporated; member, National Science Board

Ruth L. Greenstein–Vice President and Treasurer, Genex Corporation; former Associate/Deputy General Counsel, National Science Foundation

B.R. Inman–Chairman and Chief Executive Officer, Westmark Systems, Inc.; former Director, National Security Agency; former Deputy Director, Central Intelligence Agency

Melvin R. Laird–Senior Counselor for National and International Affairs, Reader's Digest Association; former Counselor to the President for Domestic Affairs; former Secretary of Defense; former Member of Congress

Joshua Lederberg–President, Rockefeller University

Franklin A. Lindsay–Chairman, Vectron, Inc.; Chairman, National Bureau of Economic Research

John L. McLucas–President and Chairman of the Board, Questech, Inc.; former Executive Vice President and Chief Strategic Officer, COMSAT; former Administrator, Federal Aviation Administration; former Secretary of the Air Force

Richard A. Meserve–partner, Covington and Burling; former counsel to the President's Science and Technology Adviser

G. William Miller–Chairman, G. William Miller and Company, Inc.; former Chairman and Chief Executive Officer, Textron, Inc.; former Chairman, Federal Reserve Board; former Secretary of the Treasury

Leif H. Olsen–Chairman, Leif H. Olsen Associates; former Chief Economist, Citibank, Inc.

Rudolph A. Oswald–Director, Department of Economic Research, AFL-CIO; member, Advisory Committee on Trade Negotiations, Office of the U.S. Special Trade Representative

[86]National Academy of Sciences. *Balancing the National Interest*, p. x.

[87]*Ibid.*, pp. 35–36, 181–182.

[88]See National Academy of Sciences. *International Competition in Advanced Technology: Decisions for America.* Washington, DC: National Academy Press, 1983; U.S. Congress. Senate. Committee on Finance. *National Research Council Report on International Competition in Advanced Technology.* Hearing, 98th Congress, 1st Session. Washington, DC: U.S. Govt. Print. Off., 1983.

[89]Among those addressing the Allen panel at its first meeting in July 1985, was Assistant Secretary of Defense Richard N. Perle. A strong advocate of national security export controls, he pledged full cooperation with the panel. Curiously, in September 1985, Perle's close colleague, Stephen D. Bryen, Deputy Assistant Secretary of Defense for International Economic, Trade, and Security Policy, publicly complained that export controls were outside of the Academy's competence and expressed puzzlement as to why the Pentagon had provided a reported $200,000 for the Allen panel. These remarks prompted some momentary concern that Bryen might prevail upon Perle to obstruct or halt the study. See Daniel S. Greenberg. "DoD Aide Scorns Academy's Study of Exports." *Science and Government Report*, v. 15, September 15, 1985, pp. 1–3.

In February 1986, however, some Pentagon funds for the Academy study were cut off and Bryen instructed his staff that they were not to engage in any further substantive cooperation of any kind with the Allen panel. See U.S. Congress. House. Committee on Science, Space, and Technology. *National Academy of Sciences Report on International Technology Transfer.* Hearings, 100th Congress, 1st Session. Washington, DC: U.S. Govt. Print. Off., 1987, pp. 46, 121–122.

[90]National Academy of Sciences. *Balancing the National Interest*, pp. 36–37; also see National Academy of Sciences. *Balancing the National Interest: Working Papers.* Washington, DC: National Academy Press, 1987.

[91]*Ibid.*, p. vi.

[92]*Ibid.*, p. 41.

[93]*Ibid.*, p. 51.

[94]*Ibid.*, pp. 41, 46.

[95]*Ibid.*, p. 46; see U.S. Department of Defense. Office of the Undersecretary of Defense for Policy. *Assessing the Effect of Technology Transfer on U.S./Western Security: A Defense Perspective.* February, 1985. Washington, DC.

[96]National Academy of Sciences. *Balancing the National Interest*, p. 110; for another similar, but later, critique of the Pentagon study, see Macdonald, *Technology and the Tyranny of Export Controls*, pp. 91–98.

[97]National Academy of Sciences. *Balancing the National Interest*, p. 110.

[98]*Ibid.*, pp. 49–50.

[99] *Ibid.*, p. 116.

[100]See *Ibid.*, pp. 123–124.

[101]*Ibid.*, p. 103.

[102]*Ibid.*, p. 131.

[103]See *Ibid.*, pp. 150–166.

[104]See *Ibid.*, pp. 167–177.

[105]102 Stat. 1107.

[106]See U.S. Congress. House. Committee on Foreign Affairs. *National Security Export Controls: Report by the National Academy of Sciences.* Hearing, 100th Congress, 1st Session. Washington, DC: U.S. Govt. Print. Off., 1988.

[107]See U.S. Congress. House. Committee on Science, Space, and Technology. *National Academy of Sciences Report on International Technology Transfer*, pp. 43–45.

[108]*Ibid.*, pp. 88–91.

[109]*Ibid.*, p. 102.

[110] *Ibid.*, p. 107.

[111]U.S. Congress. House. Committee on Science, Space, and Technology. *Export Controls, Competitiveness and International Cooperation: A Critical Review.* Committee print, 101st Congress, 1st Session. Washington, DC: U.S. Govt. Print. Off., 1989, p. ix.

[112]See *Ibid.*, pp. 12–14, 37–40.

[113]*Ibid.*, p. 13.

[114]Macdonald, *Technology and the Tyranny of Export Controls*, p. 41.

[115]As identified in its report, the members of the study committee were as follows:

Seymour E. Goodman, Chairman – University of Arizona
Duane A. Adams – Carnegie-Mellon University
Isaac L. Auerbach – Auerbach Consultants
Warren Davis – Semiconductor Industry Association
Jack J. Dongarra – Argonne National Laboratory
Thomas F. Gannon – Digital Equipment Corporation
David C. Gossard – Massachusetts Institute of Technology
Anthony C. Hearn – RAND Corporation
John L. Hennessy – Stanford University
John McCarthy – Stanford University
William K. McHenry – Georgetown University
John K. Ousterhout – University of California at Berkeley
Theodore J. Ralston – Microelectronics and Computer Technology Corporation
James W. Stansberry – Stansberry Associates, Inc.
Judith Thornton – University of Washington
Clark Weissman – Unisys Corporation
Troy L. Wilson – IBM Corporation

[116]National Research Council. *Global Trends in Computer Technology and Their Impact on Export Control.* Washington, DC: National Academy Press, 1988, pp. 1–3.

[117]See *Ibid.*, pp. 218–242.

[118]102 Stat. 1107 at 1363.

[119]National Academy of Sciences. *Finding Common Ground.* Washington, DC: National Academy Press, 1991, p. 7; the COSEPUP charge to the study panel may be found in *Ibid.*, pp. 306–307.

[120]As identified in its report, the members of the study panel were as follows:

Roland W. Schmitt, Chairman – President, Rensselaer Polytechnic Institute; member, National Science Board of the National Science Foundation; former chairman, National Science Board (1984–1988)

William F. Burns, Vice Chairman – Major General (Ret.), U.S. Army; former Director, U.S. Arms Control and Disarmament Agency (1988); former Principal Deputy Assistant Secretary of State, Bureau of Politico-Military Affairs (1986–1988)

Arden L. Bement, Jr. – Vice President, Technical Resources, TRW; member, National Science Board of the National Science Foundation; former Deputy Under Secretary of Defense for Research and Engineering (1979–1981)

Ashton B. Carter—Professor of Public Policy, John F. Kennedy School of Government; Director, Center for Science and International Affairs, Harvard University

Kenneth W. Dam—Vice President, Law and External Relations, IBM Corporation; former Deputy Secretary of State (1982–1985)

Herbert M. Dwight, Jr.—President, Superconductor Technologies, Inc.

John L. Ellicott—Chairman, Management Committee, Covington and Burling

Lincoln D. Faurer—Lieutenant General (Ret.), U.S. Air Force; President and Chief Executive Officer, Corporation for Open Systems; former Director, National Security Agency (1981–1985)

Charles Gati—Professor of Political Science, Union College

Seymour E. Goodman—Professor of Management Information Systems and Policy, College of Business and Public Administration, University of Arizona

Ruth L. Greenstein—Vice President, Administration and Finance, Institute for Defense Analysis; former Vice President and General Counsel, Genex Corporation (1984–1990); former Associate/Deputy General Counsel, National Science Foundation (1981–1984)

Benjamin Huberman—President, Consultants International Group, Inc.; former Deputy Director, Office of Science and Technology Policy (1978–1981); former senior staff member, National Security Council (1978–1981)

Raymond Kline—President, National Academy of Public Administration; former Associate Administrator for Management Operations, National Aeronautics and Space Administration (1977–1979)

Robert Legvold—Director, W. Averell Harriman Institute for Advanced Study of the Soviet Union, Columbia University

Boyd J. McKelvain—Manager, Corporate Export Administration Operation, General Electric Company; Chairman, Technical Advisory Committee on Implementation of the Militarily Critical Technologies List, Department of Commerce

John L. McLucas—Aerospace consultant; former Secretary of the Air Force (1972–1975)

M. Granger Morgan—Head, Department of Engineering and Public Policy, and Professor of Engineering and Public Policy/Electrical and Computer Engineering, Carnegie Mellon

William J. Perry—Chairman, Technology Strategies and Alliances; Co-Director, Center for International Security and Arms Control, Stanford University; member, Defense Science Board; former Under Secretary of Defense for Research and Engineering (1977–1981)

O.M. Roetman—Vice President, Government and International Affairs, Boeing Company

Gaston J. Sigur—Distinguished Professor of East Asian Studies, School of International Affairs, George Washington University; former Assistant Secretary of State for East Asian and Pacific Affairs (1986–1989); former Special Assistant to the President for National Security Affairs (1982–1984)

John D. Steinbruner—Director of Foreign Policy Studies, Brookings Institution

Paula Stern—President, The Stern Group; former Chair, U.S. International Trade Commission (1984–1986)

[121]National Academy of Sciences, *Finding Common Ground*, pp. 8–9; subpanels and their members may be found in *Ibid.*, pp. v-vi; subpanel reports and panel staff's research papers are included in appendices of the final report.

[122]When mandating the study in August 1988, the 100th Congress clearly anticipated its timely completion for the reauthorization of the Export Administration Act by the next Congress. Indeed, the automatic termination date for the act was moved ahead 1 year to September 30, 1990 (see 99 Stat. 155; 102 Stat. 1362). Efforts at producing the study, however, got a late start. After a long struggle in 1990, the 101st Congress in its closing days reached agreement on a reauthorization measure, but it was vetoed after the final

adjournment. There was little enthusiasm for the Schmitt panel's report on Capitol Hill when it appeared shortly after the convening of the 102nd Congress.

[123]National Academy of Sciences, *Finding Common Ground*, pp. 165–166.

[124]*Ibid.*, p. 168.

[125]*Ibid.*

[126]*Ibid.*, pp. 171–174.

[127]See *Ibid.*, pp. 178–180.

[128]*Ibid.*, p. 181.

[129]*Ibid.*, pp. 182–183.

[130]*Ibid.*, p. 183.

[131]*Ibid.*, pp. 187–190.

[132]*Ibid.*, pp. 190–195.

[133]*Ibid.*, pp. 39–40.

[134]See *Ibid.*, pp. 12–17, 106–117.

[135]See *Ibid.*, pp. 40–42.

[136]Harold D. Lasswell. *National Security and Individual Freedom.* New York: McGraw-Hill Book Company, 1950, p. 56.

[137]See National Academy of Sciences, *Finding Common Ground*, pp. 187–191.

6

THE POST-COLD WAR SETTING

With the succession of Mikhail Gorbachev to the position of General Secretary of the Communist Party in March 1985, a tide of change began to arise modestly in the Soviet Union and slowly sweep across that nation and Eastern Europe. It was a tide that produced dramatic events and one that perhaps has not yet come to rest. It has had, of course, an impact on the changing fortunes of American national security in matters of the diplomacy, defense, economics, and politics of the United States. Furthermore, it will, in all likelihood, continue to do so in the foreseeable future.

This development, although certainly of considerable significance, is not the only one to have seriously affected American national security as the final decade of the 20th century elapses. Among the many who are unhappy with the international trade position and trade deficit of the United States, strong resentments exist toward certain Pacific rim nations—particularly Japan—thought to be selfishly exploiting almost every opportunity in the American open market for their own advantage and sometimes engaging in unfair trade practices, including industrial espionage.[1] In fact, with the collapse of the Soviet Union and its military threat, the United States, in the view of some, no longer really has a reason to put strategic interests ahead of commercial ones in promoting a strong Japanese economy. Both this trade situation and the concomitant trade position of Japan have contributed to the sense of a security threat. By the early weeks of 1993, Japan's trade surplus was reportedly a record $107 billion, the first time any nation's exports exceeded its imports by more than $100 million.[2]

Within this evolving climate of opinion, congressional dissatisfaction with the availability of translated Japanese scientific and technical information in the United States resulted in the establishment of the Japanese technical literature program in the Department of Commerce in 1986.[3]

Congressional concern and unhappiness with some of the trading practices of Japanese industries and those of other Pacific rim nations were also reflected in the Omnibus Trade and Competitiveness Act of 1988. The statute was largely generic in character—not country, industry, or company specific—and tended to refine, rather than revolutionize, existing trade policy and policymaking procedures. Furthermore, a proposal by Representative Richard Gephardt, providing that failure to negotiate the elimination of a country's unfair trading practices would require the President to take measures to achieve annual 10% cuts in that country's trade surplus with the United States, was dropped. Substituted for this provision was one requiring the U.S. Trade Representative to identify "priority" unfair trade practices as well as "priority" foreign trade partners following such practices. This having been done, the Trade Representative was to begin investigations of such practices, negotiate for up to three years with each foreign country engaging in them, and retaliate if necessary.[4]

Another provision made it easier for American firms to obtain remedies against pirated products imported into the U.S., defined the lack of intellectual property rights protection in international trade as an unfair trade practice, and mandated expedited Trade Representative action against certain violators, allowed process patent holders to seek damages from violators in federal courts, and made intellectual property rights protection an objective of the Uruguay Round of multilateral trade negotiations then underway.[5]

Still another provision required the President to "impose, for a period of three years, a prohibition on contracting with, and procurement of products and services from, the Toshiba Corporation and Kongsberg Vaapenfabrikk, by any department, agency, or instrumentality of the United States Government" as a consequence of illegal sales to the Soviet Union by these firms.[6]

Finally, the new trade law gave the Japanese concern because of its policy of unilateralism or forcing dispute settlements through unilateral U.S. action and not following multilateralism, particularly through the General Agreement on Tariffs and Trade.

In the spring of 1989, another matter—proposed joint U.S.-Japan development of a modified version of the F-16 fighter to meet special Japanese requirements—ran into congressional opposition and became a lightning rod for a bipartisan coalition angry at Japan's alleged exploitation of its commercial and military relations with the United States.[7] In mid-May, the Senate attempted to block the deal outright, but, failing that, then voted stringent conditions aimed at protecting U.S. technology and intellectual property involved in the joint venture. The House followed suit in early June. President Bush subsequently vetoed these modifications and was sustained in the Senate, but the underlying issue was not settled. Soon, a broad range of Japanese trade practices were no longer merely irritating, but coming to be viewed as threatening to the national security of the United States. By late April 1993, both the United States and Japan had adopted tough new trade positions reflective of mounting economic friction between the two countries.[8]

The national security concept, indeed, was starting to undergo a revision in terms of changing international conditions, perhaps a more profound change than at any previous time. An old nemesis, one very much a part of the definition of national security, was disappearing. New threats, however, seemed to be emerging to take its place.

THE SOVIET COLLAPSE

During his long leadership of the Soviet Union, Leonid Brezhnev—General Secretary of the Communist Party since October 1964 and chairman of the Presidium of the U.S.S.R. Supreme Soviet since June 1977—presided over a steady increase in the living standards of his country and an expansion of its influence throughout the world. However, it was also a dogmatic, authoritarian, repressive, and cheerless regime composed of aging bureaucrats. Brezhnev himself began to experience serious ill health at the end of the 1960s, and suffered a major stroke in January 1976, from which he never fully recovered. He remained in authority for six more years and then, just as doubt about his continued physical and political power began to grow in the West, he died suddenly on November 10, 1982. An unusually quick leadership transition followed. Two days after Brezhnev's death and three days before his burial, Yuri Andropov, the head of the KGB or secret police, became General Secretary of the Communist Party. In June of the following year, he was elected to the chairmanship of the Presidium of the Supreme Soviet.

Something of a reformer, Andropov, 68 years old, also was in poor health. However, he had a disciple of sorts—Mikhail Gorbachev. Once again, leadership suddenly changed. Andropov died and four days later Konstantin Chernenko was named General Secretary on February 13, 1984. A Brezhnev follower of the old school and vintage, Chernenko was last seen in public at the end of December 1984. Death came on the evening of March 10, 1985. The next day, an extraordinary session of the Central Committee named Mikail Gorbachev its General Secretary–the third occupant of the office in three and one-half years. Only 54 years old, Gorbachev pursued a number of reforms initiated by Andropov and would soon reveal a revitalization program of his own.[9]

By the end of 1989, after about four years, Gorbachev had created a younger, somewhat pluralistic, better educated, leadership of largely his own choosing or approval.[10] A reform program of *perestroika* or restructuring, which had *glasnost* or openness as one of its central tenets, began to be revealed at Central Committee meetings in January and June 1987.[11] At the latter session, there was agreement to Gorbachev's proposal for a party conference–the first such gathering in nearly 50 years–in the summer of 1988 to consider various so-called democratizing measures. This was the 19th party conference of June 28 to July 1, 1988. A remarkable meeting, it produced and adopted a series of resolutions calling for further democratization of Soviet society and politics. Endorsed at Central Committee meetings in July and September, these proposals resulted in constitutional reforms approved in November and December of 1988–a new state structure with a working parliament, a constitutional review commission, and reforms in electoral law.[12] The parliamentary plan received fulfillment in March 1989, with elections to the new Congress of People's Deputies, which saw the defeat of various prominent Communist Party candidates.[13]

The implementation of *perestroika* also meant that Moscow was giving up its domination of Eastern Europe.[14] In Hungary, long-time Communist Party leader Janos Kadar resigned as Secretary General in late May 1988 and subsequently died a year later on July 6, 1989. By then, Communist Party leaders, pressured for reforms, had signed an accord with opposition groups on June 10 to negotiate for a multiparty system. Shortly thereafter, on June 24, a four-man presidency was elected, including economic reformers. In early October, the Communist Party formally disbanded and took its place in a multiparty system as a new Socialist Party.

In early September 1989, Hungary opened its border with Austria, initially allowing some 30,000 East Germans to flee their country with more to follow in subsequent weeks. Erich Honecker, stalwart leader of the Communist Party of the German Democratic Republic, was replaced by Egon Krenz in mid-October. November saw a number of dramatic events. Czechoslovakia opened its border with Austria on November 3, permitting more East Germans to escape to the West. A massive demonstration for democratic reforms occurred the next day in East Germany, followed by the resignation of the government on November 7, the resignation of almost half the Politburo on November 8, and the toppling of the Berlin Wall on November 9 when East Germany allowed unrestricted travel between its sector of Berlin and the free zone. During the first days of December, the Communist Party gave up its monopoly of political power in East Germany, the rest of the Politburo and the entire Central Committee resigned, and Honecker and other members of the old leadership were placed under house arrest on suspicion of corruption. International and inter-German negotiations on reunification began early the following year, were successful, and the German Democratic Republic ceased to exist in October 1990.

Elsewhere, in Czechoslovakia, massive November 1989 demonstrations in Prague for free elections and replacement of the government prompted the resignation of the Communist Party leadership and the allowance of noncommunist parties that would have access to the news media. By the end of the year, Czechoslovakia had opened its border with Austria to allow free movement to the West, instituted various democratic reforms, and made Vaclav Havel, a leading opposition figure, its president.

A number of democratic reforms occurred in Poland following the national elections of June 4, 1989, which routed the Communists. There was a far different experience, however, in Romania. Demands for reform resulted in widespread riots and the killing of state security forces in Bucharest during December 1989. The success of the revolt was assured when army troops refused to fire on their countrymen. Although President Nicolae Ceausescu and his wife attempted to flee, they were captured, tried in secret, and executed on Christmas Day.

Democratic reforms continued in the Soviet Union. The new parliamentary system aided the process with new laws openly debated, freely reported, and accessible to all. In February 1990, the Central Committee accepted Gorbachev's proposal, among others, that the Communist Party should abandon its monopolistic position and that an executive presidency with enlarged powers should be created.[15] Both ideas were subsequently given constitutional approval, but in a late 1990 retrenchment, Gorbachev seemed to back away from his reform agenda somewhat by strengthening the control of the president over central state power, rejecting a proposed 500-day plan for economic reform of the nation, and initiating new law enforcement authority that immediately had severely adverse impact upon the breakaway Baltic republics.[16] Indeed, Gorbachev was aware that his reforms not only had not gone well for all quarters of Soviet society, but also that some within the old power structure were unhappy with the new regime he was creating. Their displeasure was revealed in August 1991 when a sudden coup occurred during Gorbachev's absence from Moscow. Within about 48 hours, however, it became apparent that this reactive revolt had little or no popular support. Widespread opposition was rallied by Boris Yeltsin, a radical progressive who had been popularly elected to the presidency of the Russian republic. Gorbachev was soon returned to the capital and reassumed his position, but now Yeltsin had the support of public opinion. The political coherency of the Soviet Union was no longer viable; the various republics were seeking independent sovereignty. All efforts at reconciliation failed. In early December, as the leaders of Byelorussia, Russia, and the Ukraine abandoned the old union and signed a declaration of a new Slavic commonwealth, Secretary of State James Baker observed that "the Soviet Union as we've known it no longer exists."[17] By the end of the year, Gorbachev resigned his presidency and, by parliamentary action, the Union of Soviet Socialist Republics formally and officially dissolved.

THE AMERICAN REACTION

During most of 1989, President Bush exhibited a very cautious approach to the changing Soviet circumstances. Statements by the new Secretary of Defense, Dick Cheney, indicated that he had little confidence in Gorbachev's reforms and thought that the Soviet Union would readily revert to its previous authoritarian ways. Similar viewpoints were expressed by Vice President Dan Quayle.[18] Eventually, as the Gorbachev reforms took root and the East bloc nations replaced their communist leaders and embarked on democratic practices, the Bush Administration assumed a more positive and cooperative attitude toward the Soviet Union. Accordingly, the 1991 edition of *Soviet Military Power* appeared in a toned-down, abbreviated version entitled *Military Forces in Transition*. "Because of the profound uncertainty on so many matters," Secretary Cheney, ever cautious, wrote in his introduction, "we have purposely not discussed a variety of issues."[19] No less a critic than former President Richard Nixon complained as late as March 1992 about the Bush Administration's reticence regarding the former Soviet Union and "pathetically inadequate response . . . to help Russia's first democratic, free-market oriented, non-expansionist government."[20]

It was only after the popularly rejected August 1991 attempt to topple Gorbachev, in which electronic communication technology played an important if not decisive role, that the Bush Administration, pressured by Congress, American business, and European allies, agreed to reexamine export curbs on selective high-technology equipment.[21] When American Telephone and Telegraph vice president Carleton S. Fiorina told a House subcommittee in late September that U.S. export restrictions on the sale of advanced telecommunications equipment were not only adversely affecting American business sales to the U.S.S.R., but also hindering Soviet progress toward democratic reforms and a free-market economy, administration witnesses rejoined that national security required such restrictions. Indeed, at issue, said Deputy Assistant Secretary of State Christopher G. Hankin was the ability of the National Security Agency to eavesdrop more easily on the current, antiquated Soviet telecommunications system than it could if fiber-optic cable was being used—the first such public admission by an administration official.[22] In March 1992, the United States finally agreed with its major allies to allow the former Soviet republics to purchase certain state of the art telecommunications equipment, including fiber-optic transmission lines.[23]

Elsewhere, at an October 1991 conference on American international business security, keynote speaker Michelle Van Cleave, assistant national security policy director for the President's Office of Science and Technology Policy, reportedly chided American businesses for putting too little emphasis on recognizing the dangers of international industrial espionage. Earlier, Stanislav Levchenko, a former Soviet KGB officer who had defected in 1979, warned conferees that the deteriorating Soviet economy would prompt Moscow to increase its industrial espionage activities in the years ahead. Support for this viewpoint was offered the following day by Harry B. Brandon, Deputy Assistant Director for Intelligence at the FBI. He told his conference audience that, where espionage operations by Czechoslovakia, Hungary, and Poland in the United States had "shut down" and those of East Germany had ceased with the demise of that country, the Soviet Union, by contrast, was greatly increasing its espionage efforts to gain economic advantage and recoup from the loss of its satellites in Eastern Europe.[24]

NEW AND RENEWED THREATS

Such visions and warnings of a continued Soviet effort to threaten U.S. national security through the unauthorized acquisition of American scientific and technological knowledge offered little justification for the application of national security controls to other than weapons and implements of war. Yet, the body of national security law that had been applied during the Cold War to restrict traditional scientific and technological communication in the past remained available for use. If old enemies faded or disappeared, perhaps new ones could be readily found.

Functionally, *enemies*—the term may be synonymous with or include attackers, nemeses, foes, opponents, competitors, adversaries, and antagonists—symbolize the antithesis of our core values and beliefs, also offering a contrast by which to measure or inflate our own worth and values. They also help identify sources of frustration, serve as scapegoats and provide a means of diverting attention from other pressing and more difficult problems, and justify actions of questionable propriety or legality. The identification and combatting of enemies affects social and political structure and goals, may foster orthodoxy, and serve as a measure of individual conformity.[25] Certainly the Soviet Union was of such significance and importance for the United States during the past decades.[26] Singing about all the things that are "wrong with these kids today" in the 1960 musical *Bye, Bye, Birdie,* a parent plaintively blurts out, "How're we gonna beat the Russians?"

New enemies, threatening national security, do appear to be in the making, as recent congressional debate on defense spending well illustrates.[27] For some, with the collapse of the Soviet Union, the strategic positions of Japan and Western Europe—economically and politically—are losing some of their past importance, while the international trade position of the former and the potential of the European Community are being seen as dangerous for U.S. economic interests.[28] In late September 1991, the General Accounting Office, for example, reported that a survey of 61 U.S. companies and government laboratories revealed 59 instances, disclosed by the respondents, in which Japanese firms withheld advanced equipment or critical materials from them or postponed delivery for at least six months while making the same goods available sooner to their own U.S. affiliates.[29] Such actions are not without significance for competitiveness and, in the minds of some, are perhaps worthy of imitation by American firms.

In Europe, the dismantling and dispersing of Soviet armed forces, for some, began to reduce the importance of NATO—fundamentally, always a military alliance—at a time when actual scaling down of the organization could weaken U.S. influence with its old allies and at a time of uncertainty about trade relationships in the face of a rising European Community. At home, an array of American politicians, diplomats, and academicians warned that a rising tide of isolationism in the country was prompting the public to call for a tougher stance on trade, to demand further U.S. troop reductions in Europe, and to regard NATO as no longer necessary, relevant, or affordable.[30]

In late November 1991, John P. Cregan, president of the 1,500-member United States Business and Industrial Council, urged fellow conservatives to abandon their traditional commitment to free trade and support a foreign economic policy that "puts American interests first in responding to a world with new economic superpowers and superpower wannabes."[31] Such a foreign economic policy might similarly restrain the availability of advanced U.S. scientific and technological knowledge for reasons of national security.

The collapse of the Soviet Union left many of its scientists without income or a livelihood.[32] In testimony before the Senate Committee on Governmental Affairs in early 1992, Director of Central Intelligence Robert M. Gates expressed concern that the emigration of these scientists to other nations could result in new challenges to American national security. According to Gates, "Cuba, India, Syria, Egypt, and Algeria are most likely to have the contacts and resident scientists to assist emigrating Soviets."[33] Apart from developing nuclear, chemical, and biological weapons, these emigrating Soviet scientists might also assist with converting dual use technology to purposes inimical to U.S. national security. In a discussion of this concern a few days later at a Senate Committee on Armed Services hearing, Senator Sam Nunn suggested that the Bush Administration might consider paying Soviet nuclear scientists to stay in their own country and assist with storing and dismantling the nuclear weapons arsenal inherited by the independent republics.[34] Secretary of State Baker subsequently announced that the United States, Germany, and Russia were planning to establish a science and technology policy institute to employ former Soviet scientists in an attempt to reduce the possibility of them selling their services to other countries.[35] American entrepreneurs also evidenced interest in recruiting Soviet scientists.[36] Meanwhile, some in the American scientific profession sought to instill values of scientific freedom and responsibility among scientists of Eastern Europe and the Russian Federation.[37]

At the end of January 1992, the chief of the armed forces of the new Commonwealth of Independent States succeeding the U.S.S.R. announced that steps were being taken to reduce arms exports and weapons leakages to Iran and other "extremist states." The emigration of Soviet scientists to these nations was not addressed.[38] However, Iran, indeed, was coming to be seen by U.S. officials as a serious national security threat.[39] Furthermore, although several of the non-Russian industrialized republics of the former Soviet Union had pledged to regulate the availability of nuclear weapons technology within their borders "in the shortest possible time," effective export controls were still not in place in the spring of 1993, giving many aspiring nuclear nations, such as Iran, ample opportunity to arm themselves with such weapons.[40]

Nearby, in the aftermath of the Persian Gulf conflict, Iraq was less than cooperative about United Nations inspections to destroy nuclear and chemical weapons. In early February 1992, Baghdad announced it would no longer accept future U.N. monitoring of its arms building or production, a violation of the Persian Gulf cease-fire agreement.[41] By the end of the month, Iraq was refusing to allow a U.N. team to begin destroying equipment used for ballistic missile construction.[42] Certainly there are other countries that, as producers and merchants of weapons of mass destruction and as abettors of and contributors to acts of terrorism, constitute a national security threat to the United States.

Finally, cutbacks in U.S. armed forces manpower are prompting increased reliance on technology to maintain the American defense posture.[43] It is likely that the Pentagon, in view of this trend, will be more sensitive to protecting American scientific and technological knowledge having potential military application—including, certainly, dual use varieties—or otherwise having a potential detrimental effect on critical technology it is using or is about to utilize. Such a strategy seemingly would also be consistent with Pentagon plans for preserving American global military supremacy and thwarting the emergence of a rival superpower anywhere in the world.[44]

NEW POLICY

There are, indeed, various new and renewed threats to America that could reinvigorate resort to the body of national security law that has been used to restrict traditional scientific and technological communication in the past. In some cases, conditions may warrant such protective action.[45] The controversy surrounding the use of this authority of late, however, strongly suggests that some reconsideration is merited.

Opportunities and vehicles for such reconsideration are being offered in the context of post-Cold War adjustments. Shortly after the convening of the 103rd Congress, for example, Senator Daniel P. Moynihan proposed the creation of a bipartisan national commission to study and recommend ways to overhaul the security classification system, particularly the elimination of the classification of nonsensitive information.[46] Shortly thereafter, congressional managers agreed to reauthorize the Export Administration Act temporarily, without modification, until the end of June 1994. Prior to that date, plans call for the statute to be completely rewritten to reflect changed world conditions.[47] In May, President Clinton announced the creation of an interagency task force to make changes in the security classification system and by the following month, in pursuit of this mission, the panel scheduled public hearings to received reform proposals and suggestions.[48]

Short of having a comprehensive vehicle—such as Senator Moynihan's proposed national commission or President Clinton's Task Force on National Performance Review, which evaluated government operations and programs during the spring and summer of 1993—for effecting reform, an authority-by-authority approach might be pursued. Policymakers could be apprised of conditions meriting examination and reform or might be acquainted with particular reform proposals. Opportunities in this regard also may emerge from presidential initiatives, such as President Clinton's interagency committee on security classification reform, or the congressional reauthorization process, such as the case of the Export Administration Act.

How might reform proceed in these situations? Taking the example of the reauthorization of the Export Administration Act, renewal of the statute will probably occur in an atmosphere of review and assessment of changed world conditions and their significance for U.S. export policy. To this discussion could be brought arguments for and examples of the importance of scientific and technological achievement—and how that achievement is fostered by a tradition of professional communication—for both the security and economic well-being of the nation. Legislative recognition of this value might begin with the declaration of policy proposed for the Export Administration Act by Representative Hamilton in 1983 or similar language supporting traditional scientific communication, with accompanying statements in the committee report expressing legislative intent that such communication should be presumptively unfettered.[49]

Another approach might be to amend the Export Administration Act in accordance with recommendations offered by the National Academy of Science's Panel on Scientific Communication and National Security. The Panel recommended prohibiting restrictions of any kind limiting access to or communication of university research, whether basic or applied, unless it involved a technology that:

- Is developing rapidly, and the time from basic science to application is short.
- Has identifiable direct military applications, or is dual use and involves process or production-related techniques.
- If transferred to a major hostile power, such as, in the past, the former Soviet Union, would give that adversary a significant, near-term military advantage.
- Is one for which the United States, or other friendly nations having control systems equal to those of the United States, are the only source of information about it.[50]

In addition, the panel proposed that unclassified scientific and technological information that is available domestically should receive a general license, that is, an exemption, from the formal Export Administration Act licensing process. It further recommended "that information that is not directly and significantly connected with technology critical to the national security" also receive a general license.[51]

Another legislative strategy—the converse of the approach taken by the Panel on Scientific Communication and National Security in the recommendations mentioned previously—might be to establish statutory criteria that must be met in order that national security export controls might be imposed on traditional scientific communication. Such criteria would constitute a definition of "national security" in this regard.

Finally, consideration might be given to modifying the term *technology* as used in the Export Administration Act so as to exclude university research results not meeting the criteria for protection specified by the Academy Panel on Scientific Communication and National Security.[52] The committee report on such a statutory redefinition could explain the legislative intent underlying the change and address, as well, the need for adjustment of the comparable term *technical data* in export regulations implementing the Export Administration Act.[53]

Alternatively, a categorical approach to reforming the imposition of statutory national security controls on traditional scientific communication might be considered. Congress could legislate statutory criteria that must be met in order that any national security controls, in other than a time of formally declared war, might be imposed on traditional scientific communications.[54] A presidential declaration on or certification of the satisfaction of the criteria and consequent imposition of such controls might be required.[55] If thought necessary, provision also could be made that, in the event that majorities in both Houses of Congress disagreed with such a declaration, it could be rescinded with a joint resolution of disapproval (which, requiring the President's signature to become effective, would have to be adopted by extraordinary majorities in each chamber to indicate that a possible veto could be overridden).[56]

Another categorical approach, generally known as "free trade in ideas," extends beyond the scientific community to restrict presidential authority to regulate the exchange of information with, travel to or from, and educational and cultural exchanges with foreign countries and foreign nationals by all American citizens.[57] Affected statutes include the International Emergency Economic Powers Act, the Export Administration Act, the Trading With the Enemy Act, and the Foreign Assistance Act of 1961.[58]

Strategies for effecting changes in national security laws used to restrict traditional scientific communication may vary. However, the opportunity to strengthen existing commercial protections for scientific and technological information as well as to build anew in this regard – the case of foreign investor controls being one example – should not be ignored.[59] Improving commercial safeguards could do much to enhance the role of American business and industry in thwarting the acquisition of science and technology by foreign interests posing a threat to the security of the United States.

In the final analysis, it is in the interest of many citizens, and not merely scientists, technologists, and their surrogates, to pursue such policy reforms in various arenas. Indeed, from the freedom of scientific communication come benefits for the entire country. Diminishment or restraint of that freedom must be addressed responsibly in open debates where opposing viewpoints may be heard and assessed by drawing on available knowledge. From such action and its contribution to policy-making, a democracy such as the United States realizes security internally as well as among the community of nations.

NOTES AND REFERENCES

[1]See Don Oberdorfer. "U.S.-Japan Relations Seen Suffering Worst Downturn in Decades." *Washington Post*, March 1, 1992, pp. A1, A26; T. R. Reid and Paul Blustein. "Japanese View U.S. With New Negativity." *Washington Post*, March 1, 1991, pp. A1, A26. Regarding industrial espionage, see Peter Schweizer. *Friendly Spies*. New York: Atlantic Monthly Press, 1993; *Congressional Record*, v. 139, April 21, 1993, pp. H1979-H1980; John Mintz. "CIA: French Targeted Secrets of U.S. Firms." *Washington Post*, April 27, 1993, pp. C1, C6; *Congressional Record*, v. 139, May 13, 1993, pp. S5966-S5969.

[2]Paul Blustein. "Japanese Trade Surplus Hits Record $107 Billion." *Washington Post*, January 23, 1993, p. G1.

[3]100 Stat. 811; see U.S. Congress. House. Committee on Science and Technology. *The Availability of Japanese Scientific and Technical Information in the United States*. Committee print, 98th Congress, 2d Session. Washington, DC: U.S. Govt. Print. Off., 1984; U.S. Congress. House. Committee on Science and Technology. *The Availability of Japanese Scientific and Technical Information in the United States*. Hearings, 98th Congress, 2d Session. Washington, DC: U.S. Govt. Print. Off., 1984.

[4]102 Stat. 1164–1179.

[5]102 Stat. 1179–1181, 1563–1570.

[6]102 Stat. 1365–1366.

[7]See U.S. Congress. House. Committee on Banking, Finance, and Urban Affairs. *Proposed Joint Development of the FSX Fighter With Japan*. Hearings, 101st Congress, 1st Session. Washington, DC: U.S. Govt. Print. Off., 1989; U.S. Congress. House. Committee on Science, Space, and Technology. *FSX*. Hearings, 101st Congress, 1st Session. Washington, DC: U.S. Govt. Print. Off., 1989.

[8]T.R. Reid and Paul Blustein. "Saying No to America." *Washington Post*, April 25, 1993, pp. A25, A28.

[9]Stephen White. *Gorbachev in Power*. Cambridge, UK: Cambridge University Press, 1990, p. 8.

[10]*Ibid.*, p. 21.

[11]*Ibid.*, pp. 23–24; see Mikail Gorbachev. *Perestroika: New Thinking for Our Country and the World.* New York: Harper and Row, 1987.

[12]See White, *Gorbachev in Power*, pp. 26–40.

[13]See *Ibid.*, pp. 40–51; for a profile of the new parliamentary structure and its members, see Dawn Mann, Robert Monyak, and Elisabeth Teague. *The Supreme Soviet: A Biographical Directory.* Washington, DC: Radio Free Europe/Radio Liberty and Center for Strategic and International Studies, 1989.

[14]For detailed accounts of these developments, see Bernard Gwertzman and Michael T. Kaufman, eds. *The Collapse of Communism.* New York: Times Books, 1990.

[15]White, *Gorbachev in Power*, p. 36.

[16]See Foreign Broadcast Information Service. "Gorbachev Alters Power Structure To Strengthen Presidency." February 20, 1991.

[17]David Hoffman. "Soviet Union as We've Known It Is Gone, Baker Says." *Washington Post*, December 9, 1991, p. A16.

[18]See text accompanying notes 102–104 in Chapter 4.

[19]See John Lancaster. "Pentagon Book a Casualty Of Soviet Rush to Reform." *Washington Post*, August 30, 1991, p. A25; John Lancaster. "Pentagon's Annual Soviet Military Book Tracks Swiftly Moving Target." *Washington Post*, October 5, 1991, p. A6.

[20]See David Hoffman. "U.S. Politics Constrains Role Abroad." *Washington Post*, March 10, 1992, p. A4; Associated Press. "President Says Nixon Critique Is 'Constructive'." *Washington Post*, March 11, 1992, p. A14; Don Oberdorfer. "Nixon Warns Bush to Aid Russia, Shun 'New Isolationism'." *Washington Post*, March 12, 1992, pp. A1, A22; Don Oberdorfer. "U.S. Planning Full Aid for Ex-Soviet States." *Washington Post*, March 20, 1992, pp. A1, A22.

[21]Evelyn Richards. "Confronting the Technology Trade Issue." *Washington Post*, August 25, 1991, pp. H1, H5; Stuart Auerbach. "High-Tech Export Curbs to Be Reexamined." *Washington Post*, October 11, 1991, pp. F1, F10.

[22]Stuart Auerbach. "AT&T: Export Curbs Hamper Soviet Reform." *Washington Post*, September 25, 1991, p. A16; see U.S. Congress. House. Committee on Foreign Affairs. *The Reauthorization of the Export Administration Act.* Hearings and mark-up, 102d Congress, 1st Session. Washington, DC: U.S. Govt. Print. Off., 1992, pp. 43–44.

[23]Stuart Auerbach. "U.S. Shifts, Agrees to Ease Fiber-Optics Export Curb." *Washington Post*, March 6, 1992, pp. D10, D12.

[24]George Lardner, Jr. "KGB Defector Predicts an Increase In Moscow's Industrial Espionage." *Washington Post*, October 29, 1991, p. A11; George Lardner, Jr. "Expert Says Soviets Are Still Spying in U.S.; Warsaw Pact Isn't." *Washington Post*, October 30, 1991, p. A6.

[25]See David J. Finlay, Ole R. Holsti, and Richard R. Fagen. *Enemies in Politics.* Chicago: Rand McNally, 1967, pp. 6–24.

[26]See, for example, Stephen J. Whitfield. *The Culture of the Cold War.* Baltimore: Johns Hopkins University Press, 1991.

[27]See Barton Gellman. "Debate on Military's Future Crystallizes Around 'Enemies Lists'." *Washington Post*, February 26, 1992, p. A20.

[28]In the spring of 1992, veteran economics columnist Hobart Rowen suggested either France or Germany could become the principal economic enemy of the United States. See Hobart Rowen. "Look to France, Germany For Economic Enemy No. 1." *Washington Post*, March 1, 1992, pp. H1, H6.

[29]Evelyn Richards. "Japanese Firms Withhold High-Tech Goods, GAO Says." *Washington Post*, September 25, 1991, pp. C1, C13; also see John Burgess. "U.S. Semiconductor Group Steps Up Its Criticism of Japan." *Washington Post*, March 12, 1992, p. A10.

[30]Marc Fisher. "Europeans Told of U.S. Isolationism." *Washington Post*, February 10, 1992, pp. A1, A23; also see David Hoffman. "U.S. Politics Constrains Role Abroad."

Washington Post, March 10, 1992, pp. A1, A4.

[31]Stuart Auerbach. "Conservative Group Attacks Free Trade." *Washington Post*, November 27, 1991, pp. C1, C3.

[32]See Fred Hiatt. "Russian Nuclear Scientists Seek Business, Food." *Washington Post*, January 18, 1992, pp. A1, A20.

[33]R. Jeffrey Smith. "Gates Fears Soviet 'Brain Drain'." *Washington Post*, January 16, 1992, p. A22.

[34]George Lardner, Jr. "Republics' Procurement of Arms Said to Plunge." *Washington Post*, January 23, 1992, p. A20.

[35]David Hoffman. "Atom Scientists At Ex-Soviet Lab Seek Help." *Washington Post*, February 15, 1992, pp. A1, A32-A33; David Hoffman. "Ex-Soviet Scientists To Get Aid." *Washington Post*, February 18, 1992, pp. A1, A10; Thomas W. Lippman. "Plan to Aid Russian Scientists Falters." *Washington Post*, March 25, 1992, p. A27.

[36]See John Burgess. "The Search for Soviet Science Is Well on Its Way." *Washington Post*, February 23, 1992, pp. H1, H6.

[37]See Jane Cave and Mark S. Frankel. *Setting New Ground Rules for Scientific Freedom and Responsibility in East-Central Europe and the Russian Federation*. Washington, DC: American Association for the Advancement of Science, 1992.

[38]R. Jeffrey Smith and Michael Getler. "Commonwealth To Curb Arms Exports to Iran." *Washington Post*, February 1, 1992, pp. A1, A18; also see Reuter. "Ex-Soviets Reported Seeking to Sell A-Bomb Materials." *Washington Post*, March 11, 1992, p. A16.

[39]See David Hoffman. "Iran's Rebuilding Seen as Challenge to West." *Washington Post*, February 2, 1992, pp. A1, A25.

[40]See Steve Coll. "Nuclear Goods Traded In Post-Soviet Bazaar." *Washington Post*, May 15, 1993, pp. A1, A18.

[41]Associated Press. "Iraq Defies U.N. Controls On Arms." *Washington Post*, February 5, 1992, p. A27.

[42]Trevor Rowe. "Baghdad Defies U.N. Deadline." *Washington Post*, February 29, 1992, pp. A1, A18.

[43]See John Lancaster. "Arms Research Booms As Production Lines Halt." *Washington Post*, December 10, 1991, pp. A1, A12.

[44]See Barton Gellman. "Keeping the U.S. First." *Washington Post*, March 11, 1992, pp. A1, A4.

[45]For a view of the situation during the early months of 1993, see Peter Behr. "AT&T Falls Victim to Cold War Trade Limits." *Washington Post*, August 6, 1993, pp. G1, G4; Stephen Budiansky. "Keeping Research Under Wraps." *U.S. News & World Report*, v. 114, March 22, 1993, pp. 48–50; Shawn P. McCarthy. "High-tech Partnerships Between U.S. and Russian Groups Still Are Clouded by Concerns Over Our National Security." *Government Computer News*, v. 12, June 7, 1993, pp. 50–51.

[46]See *Congressional Record*, v. 139, January 21, 1993, pp. S596-S598; of related interest, see U.S. Congress. House. Committee on Government Operations. *Government Secrecy After the Cold War*. Hearing, 102nd Congress, 2d Session. Washington, DC: U.S. Govt. Print. Off., 1993.

[47]See *Congressional Record*, v. 139, February 16, 1993, pp. H605-H609; *Ibid.*, March 11, 1993, pp. S2770-S2775. See 107 Stat. 40.

[48]See Tim Weiner. "President Moves to Release Classified U.S. Documents." *New York Times*, May 5, 1993, p. A18; *Federal Register*, v. 58, May 20, 1993, p. 29480.

[49]See U.S. Congress. House. Committee on Foreign Affairs. *Extension and Revision of the Export Administration Act of 1979*. Hearings and mark-up, 98th Congress, 1st Session. Washington,DC: U.S. Govt. Print. Off., 1985, pp. 1259–1263.

[50]National Academy of Sciences, *Scientific Communication and National Security*. Washington, DC: National Academy Press, 1982, p. 49.

[51]*Ibid.*, p. 54.

[52]See 50 U.S.C. App. 2415(4).

[53]See 15 C.F.R. 779.1 (1992).

[54]These would be, at a minimum, the national security controls established by the Export Administration Act, the Invention Secrecy Act, and the Atomic Energy Act.

[55]A model for such an arrangement may be found in the National Emergencies Act, 50 U.S.C. 1601–1651.

[56]Alternatively, before imposing controls, the President might be required to transmit his declaration to Congress and wait for a period of time before proceeding in order that objections might be expressed and discussed. This report and wait arrangement, although fostering consultation and comity with Congress, arguably might not prove to be an expedient procedure for dealing with a threatening emergency.

[57]See American Civil Liberties Union. *Free Trade In Ideas: A Constitutional Imperative.* Washington, DC: Center for National Security Studies, 1984; U.S. Congress. House. Committee on the Judiciary. *Free Trade In Ideas.* Hearings, 101st Congress, 1st Session. Washington, DC: U.S. Govt. Print. Off., 1989.

[58]See H.R. 1579, the Free Trade in Ideas Act of 1993, introduced April 1, 1993, by Representative Howard L. Berman with 14 initial co-sponsors. The bill was referred to the Committee on Foreign Affairs.

[59]See, for example, Susan J. Tolchin. "Halting the Erosion of America's Critical Assets." *Issues in Science and Technology*, v. 9, Spring, 1993, pp. 65–72; U.S. Congress. Joint Economic Committee. *Foreign Direct Investment.* Hearing, 102nd Congress, 2d Session. Washington, DC: U.S. Govt. Print. Off., 1993.

APPENDIX A

National Security Decision Directive 145: "National Policy on Telecommunications and Automated Information Systems Security"

September 17, 1984
Unclassified

Recent advances in microelectronics technology have stimulated an unprecedented growth in the supply of telecommunications and information processing services within government and throughout the private sector. As new technologies have been applied, traditional distinctions between telecommunications and automated information systems have begun to disappear. Although this trend promises greatly improved efficiency and effectiveness, it also poses significant security challenges. Telecommunications and automated information processing systems are highly susceptible to interception, unauthorized electronic access, and related forms of technical exploitation, as well as other dimensions of the hostile intelligence threat. The technology to exploit these electronic systems is widespread and is used extensively by foreign nations and can be employed, as well, by terrorist groups and criminal elements. Government systems as well as those which process the private or proprietary information of US persons and businesses can become targets for foreign exploitation.

Within the government these systems process and communicate classified national security information and other sensitive information concerning the vital interests of the United States. Such information, even if unclassified in isolation, often can reveal highly classified and other sensitive information when taken in aggregate. The compromise of this information, especially to hostile intelligence services, does serious damage to the United States and its national security interests. A comprehensive and coordinated approach must be taken to protect the government's telecommunications and automated information systems against current and projected threats. This approach must include mechanisms for formulating policy, for overseeing systems security resource programs, and for coordinating and executing technical activities.

This Directive: Provides initial objectives, policies, and an organizational structure to guide the conduct of national activities directed toward safeguarding systems which process or communicate sensitive information from hostile exploitation; establishes a mechanism for policy development; and assigns responsibilities for implementation. It is intended to assure full participation and cooperation among the various existing centers of technical expertise throughout the Executive Branch, to promote a coherent and coordinated defense against the hostile intelligence threat to these systems, and to foster an appropriate partnership between government and the private sector in attaining these goals. This Directive specifically recognizes the special requirements for protection of intelligence sources and methods. It is intended that the mechanisms established by this Directive will initially focus on those automated information systems which are connected to telecommunications transmission systems.

1. *Objectives.* Security is a vital element of the operational effectiveness of the national security activities of the government and of military combat readiness. Assuring the security of telecommunications and automated information systems which process and communicate classified national security information, and other sensitive government national security information, and offering assistance in the protection of certain private sector information are key national responsibilities. I, therefore, direct that the government's capabilities for securing telecommunications and automated information systems against technical exploitation threats be maintained or improved to provide for:

a. A reliable and continuing capability to assess threats and vulnerabilities, and to implement appropriate, effective countermeasures.

b. A superior technical base within the government to achieve this security, and support for a superior technical base within the private sector in areas which complement and enhance government capabilities.

c. A more effective application of government resources and encouragement of private sector security initiatives.

d. Support and enhancement of other policy objectives for national telecommunications and automated information systems.

2. *Policies.* In support of these objectives, the following policies are established:

a. Systems which generate, store, process, transfer or communicate classified information in electrical form shall be secured by such means as are necessary to prevent compromise or exploitation.

b. Systems handling other sensitive, but unclassified, government or government-derived information, the loss of which could adversely affect the national security interest, shall be protected in proportion to the threat of exploitation and the associated potential damage to the national security.

c. The government shall encourage, advise, and, where appropriate, assist the private sector to: identify systems which handle sensitive non-government information, the loss of which could adversely affect the national security; determine the threat to, and vulnerability of, these systems; and formulate strategies and measures for providing protection in proportion to the threat of exploitation and the associated potential damage. Information and advice from the perspective of the private sector will be sought with respect to implementation of this policy. In cases where implementation of security measures to non-governmental systems would be in the national security interest, the private sector shall be encouraged, advised, and, where appropriate, assisted in undertaking the application of such measures.

d. Efforts and programs begun under PD-24 which support these policies shall be continued.

3. *Implementation.* The Directive establishes a senior level steering group; an interagency group at the operating level; an executive agent and a national manager to implement these objectives and policies.

4. *Systems Security Steering Group.*

a. A Systems Security Steering Group consisting of the Secretary of State, the Secretary of the Treasury, the Secretary of Defense, the Attorney General, the Director of the Office of Management and Budget, the Director of Central Intelligence, and chaired by the Assistant to the President for National Security Affairs is established. The Steering Group shall:

(1) Oversee this Directive and ensure its implementation. It shall provide guidance to the Executive Agent and through him to the National Manager with respect to the activities undertaken to implement this Directive.

(2) Monitor the activities of the operating level National Telecommunications and Information Systems Security Committee and provide

guidance for its activities in accordance with the objectives and policies contained in this Directive.

(3) Review and evaluate the security status of those telecommunications and automated information systems that handle classified or sensitive government or government-derived information with respect to established objectives and priorities, and report findings and recommendations through the National Security Council to the President.

(4) Review controlled resources program and budget proposals for telecommunications systems security, including the COMSEC Resources Program, for the US Government and provide recommendations to OMB for the normal budget review process.

(5) Review in aggregate the program and budget proposals for the security of automated information systems of the departments and agencies of the government.

(6) Review and approve matters referred to it by the Executive Agent in fulfilling the responsibilities outlined in paragraph 6. below.

(7) On matters pertaining to the protection of intelligence sources and methods be guided by the policies of the Director of Central Intelligence.

(8) Interact with the Steering Group on National Security Telecommunications to ensure that the objectives and policies of this Directive and NSDD-97, National Security Telecommunications Policy, are addressed in a coordinated manner.

(9) Recommend for Presidential approval additions or revisions to this Directive as national interests may require.

(10) Identify categories of sensitive non-government information, the loss of which could adversely affect the national security interest, and recommend steps to protect such information.

b. The National Manager for Telecommunications and Information Systems Security shall function as executive secretary to the Steering Group.

5. *The National Telecommunications and Information Systems Security Committee.*

a. The National Telecommunications and Information Systems Security Committee (NTISSC) is established to operate under the direction of the Steering Group to consider technical matters and develop operating policies as necessary to implement the provisions of this Directive. The Committee shall be chaired by the Assistant Secretary of Defense (Command, Control, Communications, and Intelligence) and shall be composed of a voting representative of each member of the Steering Group and of each of the following:

The Secretary of Commerce
The Secretary of Transportation
The Secretary of Energy
Chairman, Joint Chiefs of Staff
Administrator, General Services Administration
Director, Federal Bureau of Investigation
Director, Federal Emergency Management Agency
The Chief of Staff, United States Army
The Chief of Naval Operations
The Chief of Staff, United States Air Force
Commandant, United States Marine Corps
Director, Defense Intelligence Agency
Director, National Security Agency
Manager, National Communications System

b. The Committee shall:

(1) Develop such specific operating policies, objectives, and priorities as may be required to implement this Directive.

(2) Provide telecommunication and automated information systems security guidance to the departments and agencies of the government.

(3) Submit annually to the Steering Group an evaluation of the status of national telecommunications and automated information systems security with respect to established objectives and priorities.

(4) Identify systems which handle sensitive, non-government information, the loss and exploitation of which could adversely affect the national security interest, for the purpose of encouraging, advising, and, where appropriate, assisting the private sector in applying security measures.

(5) Approve the release of sensitive systems technical security material, information, and techniques to foreign governments or international organizations with the concurrence of the Director of Central Intelligence for those activities which he manages.

(6) Establish and maintain a national system for promulgating the operating policies, directives, and guidance which may be issued pursuant to this Directive.

(7) Establish permanent and temporary subcommittees as necessary to discharge its responsibilities.

(8) Make recommendations to the Steering Group on Committee membership and establish criteria and procedures for permanent observers from other departments or agencies affected by specific matters under deliberation, who may attend meetings upon invitation of the Chairman.

(9) Interact with the National Communications System Committee of Principals established by Executive Order 12472 to ensure the coordinated execution of assigned responsibilities.

c. The Committee shall have two subcommittees, one focusing on telecommunications security and one focusing on automated information systems security. The two subcommittees shall interact closely and any recommendations concerning implementation of protective measures shall combine and coordinate both areas where appropriate, while considering any differences in the level of maturity of the technologies to support such implementation. However, the level of maturity of one technology shall not impede implementation in other areas which are deemed feasible and important.

d. The Committee shall have a permanent secretariat composed of personnel of the National Security Agency and such other personnel from departments and agencies represented on the Committee as are requested by the Chairman. The National Security Agency shall provide facilities and support as required. Other departments and agencies shall provide facilities and support as requested by the Chairman.

6. *The Executive Agent of the Government for Telecommunications and Information Systems Security.* The Secretary of Defense is the Executive Agent of the Government for Communications Security under authority of Executive Order 12333. By authority of this Directive he shall serve an expanded role as Executive Agent of the Government for Telecommunications and Automated Information Systems Security and shall be responsible for implementing, under his signature, the policies developed by the NTISSC. In this capacity he shall act in accordance with policies and procedures established by the Steering Group and the NTISSC to:

a. Ensure the development, in conjunction with NTISSC member departments and agencies, of plans and programs to fulfill the objectives of this Directive, including the development of necessary security architectures.

b. Procure for and provide to departments and agencies of the government and, where appropriate, to private institutions (including government contractors) and foreign governments, technical security material, other technical assistance, and other related services of common concern, as required to accomplish the objectives of this Directive.

c. Approve and provide minimum security standards and doctrine, consistent with provisions of the Directive.

d. Conduct, approve, or endorse research and development of

techniques and equipment for telecommunications and automated information systems security for national security information.

e. Operate, or coordinate the efforts of, government technical centers related to telecommunications and automated information systems security.

f. Review and assess for the Steering Group the proposed telecommunications systems security programs and budgets for the departments and agencies of the government for each fiscal year and recommend alternatives, where appropriate. The views of all affected departments and agencies shall be fully expressed to the Steering Group.

g. Review for the Steering Group the aggregated automated information systems security program and budget recommendations of each of the departments and agencies of the US Government for each fiscal year.

7. *The National Manager for Telecommunications Security and Automated Information Systems Security.* The Director, National Security Agency is designated the National Manager for Telecommunications and Automated Information Systems Security and is responsible to the Secretary of Defense as Executive Agent for carrying out the foregoing responsibilities. In fulfilling these responsibilities the National Manager shall have authority in the name of the Executive Agent to:

a. Examine government telecommunications systems and automated information systems and evaluate their vulnerability to hostile interception and exploitation. Any such activities, including those involving monitoring of official telecommunications, shall be conducted in strict compliance with law, Executive Orders and applicable Presidential Directives. No monitoring shall be performed without advising the heads of the agencies, departments, or services concerned.

b. Act as the government focal point for cryptography, telecommunications systems security, and automated information systems security.

c. Conduct, approve, or endorse research and development of techniques and equipment for telecommunications and automated information systems security for national security information.

d. Review and approve all standards, techniques, systems, and equipments for telecommunications and automated information systems security.

e. Conduct foreign communications security liaison, including agreements with foreign governments and with international and private organizations for telecommunications and automated information systems security, except for those foreign intelligence relationships conducted for intelligence purposes by the Director of Central Intelligence. Agreements shall be coordinated with affected departments and agencies.

f. Operate such printing and fabrication facilities as may be required to perform critical functions related to the provision of cryptographic and other technical security material or services.

g. Assess the overall security posture and disseminate information on hostile threats to telecommunications and automated information systems security.

h. Operate a central technical center to evaluate and certify the security of telecommunications systems and automated information systems.

i. Prescribe the minimum standards, methods and procedures for protecting cryptographic and other sensitive technical material, techniques, and information.

j. Review and assess annually the telecommunications systems security programs and budgets of the departments and agencies of the government, and recommend alternatives, where appropriate, for the Executive Agent and the Steering Group.

k. Review annually the aggregated automated information systems security program and budget recommendations of the departments and agencies of the US Government for the Executive Agent and the Steering Group.

l. Request from the heads of departments and agencies such information and technical support as may be needed to discharge the responsibilities assigned herein.

m. Enter into agreements for the procurement of technical security material and other equipment, and their provision to government agencies and, where appropriate, to private organizations, including government contractors, and foreign governments.

8. *The Heads of Federal Departments and Agencies shall:*

a. Be responsible for achieving and maintaining a secure position for telecommunications and automated information systems within their departments or agencies.

b. Ensure that the policies, standards and doctrines issued pursuant to this Directive are implemented within their departments or agencies.

c. Provide to the Systems Security Steering Group, the NTISSC, Executive Agent, and the National Manager, as appropriate, such information as may be required to discharge responsibilities assigned herein, consistent with relevant law, Executive Order, and Presidential Directives.

9. *Additional Responsibilities.*

a. The Secretary of Commerce, through the Director, National Bureau of Standards, shall issue for public use such Federal Information Processing Standards for the security of information in automated information systems as the Steering Group may approve. The Manager, National Communications System, through the Administrator, General Services Administration, shall develop and issue for public use such Federal Telecommunications Standards for the security of information in telecommunications systems as the National Manager may approve. Such standards, while legally applicable only to Federal Departments and Agencies, shall be structured to facilitate their adoption as voluntary American National Standards as a means of encouraging their use by the private sector.

b. The Director, Office of Management and Budget, shall:

(1) Specify data to be provided during the annual budget review by the departments and agencies on programs and budgets relating to telecommunications systems security and automated information systems security of the departments and agencies of the government.

(2) Consolidate and provide such data to the National Manager via the Executive Agent.

(3) Review for consistency with this Directive, and amend as appropriate, OMB Circular A-71 (Transmittal Memorandum No. 1), OMB Circular A-76, as amended, and other OMB policies and regulations which may pertain to the subject matter herein.

10. *Nothing in this Directive*

a. Alters the existing authorities of the Director of Central Intelligence, including his responsibility to act as Executive Agent of the Government for technical security countermeasures (TSCM).

b. Provides the NTISSC, the Executive Agent, or the National Manager authority to examine the facilities of other departments and agencies without approval of the head of such department or agency, nor to request or collect information concerning their operation for any purpose not provided for herein.

c. Amends or contravenes the provisions of existing law, Executive Orders, or Presidential Directives which pertain to the privacy aspects or financial management of automated information systems or to the administrative requirements for safeguarding such resources against fraud, abuse, and waste.

d. Is intended to establish additional review processes for the procurement of automated information processing systems.

11. For the purposes of this Directive, the following terms shall have the meanings indicated:

a. *Telecommunications* means the preparation, transmission, communication or related processing of information by electrical, electromagnetic, electromechanical, or electro-optical means.

b. *Automated Information Systems* means systems which create, prepare, or manipulate information in electronic form for purposes other than telecommunication, and includes computers, word processing systems, other electronic information handling systems, and associated equipment.

c. *Telecommunications and Automated Information Systems Security* means protection afforded to telecommunications and automated information systems, in order to prevent exploitation through interception, unauthorized electronic access, or related technical intelligence threats, and to ensure authenticity. Such protection results from the application of security measures (including cryptosecurity, transmission security, emission security, and computer security) to systems which generate, store, process, transfer, or communicate information of use to an adversary, and also includes the physical protection of sensitive technical security material and sensitive technical security information.

d. *Technical security material* means equipment, components, devices, and associated documentation or other media which pertain to cryptography, or to the securing of telecommunications and automated information systems.

13. The functions of the Interagency Group for Telecommunications Protection and the National Communications Security Committee (NCSC) as established under PD-24 are subsumed by the Systems Security Steering Group and the NTISSC, respectively. The policies established under the authority of the Interagency Group of the NCSC, which have not been superseded by this Directive, shall remain in effect until modified or rescinded by the Steering Group or the NTISSC, respectively.

14. Except for ongoing telecommunications protection activities mandated by and pursuant to PD/NSC-24, that Directive is hereby superseded and cancelled.

APPENDIX B

National Security Decision Directive 189: "National Policy on the Transfer of Scientific, Technical and Engineering Information"

September 21, 1985

I. *PURPOSE*

This directive establishes national policy for controlling the flow of science, technology. and engineering information produced in federally-funded fundamental research at colleges, universities, and laboratories. Fundamental research is defined as follows:

> "'Fundamental research' means basic and applied research in science and engineering, the results of which ordinarily are published and shared broadly within the scientific community, as distinguished from proprietary research and from industrial development, design, production, and product utilization, the results of which ordinarily are restricted for proprietary or national security reasons."

II. *BACKGROUND*

The acquisition of advanced technology from the United States by Eastern Bloc nations for the purpose of enhancing their military capabilities poses a significant threat to our national security. Intelligence studies indicate a small but significant target of the Eastern Bloc intelligence gathering effort is science and engineering research performed at universities and federal laboratories. At the same time, our leadership position in science and technology is an essential element in our economic and physical security. The strength of American science requires a research environment conducive to creativity, an environment in which the free exchange of ideas is a vital component.

In 1982, the Department of Defense and National Science Foundation sponsored a National Academy of Sciences study of the need for controls on scientific information. This study was chaired by Dr. Dale Corson, President Emeritus of Cornell University. It concluded that, while there has been a significant transfer of U.S. technology to the Soviet Union, the transfer has occurred through many routes with universities and open scientific communication of fundamental research being a minor contributor. Yet as the emerging government-university-industry partnership in research activities continues to grow, a more significant problem may well develop.

III. *POLICY*

It is the policy of this Administration that, to the maximum extent possible, the products of fundamental research remain unrestricted. It is also the policy of this Administration that, where the national security requires control, the mechanism for control of information gathered during federally-funded fundamental research in science, technology and engineering at colleges, universities and laboratories is classification. Each federal government agency is responsible for: a) determining whether classification is appropriate prior to the award of a research grant, contract, or cooperative agreement and, if so, controlling the research results through standard classification procedures; b) periodically reviewing all research grants, contracts, or cooperative agreements for potential classification. No restrictions may be placed upon the conduct or reporting of federally-funded fundamental research that has not received national security classification, except as provided in applicable U.S. Statutes.

APPENDIX C

National Security Study Directive Number 14–82: "Scientific Communication and National Security"

December 23, 1982

Introduction

This National Security Study Directive establishes the Terms of Reference for a review of scientific communication and the protection of national security information.

Objective of Review

To produce a National Security Decision Directive (NSDD), Subject: Policy for Protection of Sensitive, but Unclassified, Scientific Information.

Scope

The interagency group established below will review the issue of protecting sensitive, but unclassified scientific research information, taking into account the recommendations made by the National Academy of Sciences' Panel on Scientific Communication and National Security (the Corson Panel) in its September 30, 1982 Report. The review will result in a report on the following issues raised by the Corson Panel's Report:

- *Identification of Sensitive Scientific Information.* How can the government improve the manner in which it determines what unclassified, non-proprietary scientific research information should be subject to control so as to avoid raising fears of intrusion within the scientific research community?
- *Export Controls.* What changes, if any, are required to ensure that: (1) implementation of export control regulations does not interfere with the legitimate communication of scientific research information, and (2) the burden of compliance on the scientific community is reasonable and acceptable?
- *Contractual Controls.* When there is Federal funding of scientific research, information which has important near-term national security implications but not requiring classification can be controlled by written restrictions in the funding instrument. If such contractual restrictions are appropriate, can they replace other forms of control for all Federally-funded scientific research? Additionally, what general guidelines or appeals mechanisms are needed to assure that: (1) the restrictions are appropriate, and (2) government agencies impose procedurally and substantively compatible restrictions on the research community?
- *Visa Controls.* The use of visa restrictions and denials to prevent undesired technology loss from scientific research institutions is another possible approach. Should the visa authority be used more extensively for this purpose than at present? If so, what criteria would be appropriate and what changes in current procedures would be required?
- *Dialogue with Scientific Community.* Whatever controls are implemented, it is important for the government and the scientific community to understand more fully each other's concerns. What specific steps should be taken and what new mechanisms, if any, should be put into place to facilitate this dialogue? What form of public participation, if any, would be appropriate before implementation of any recommendations concerning the above questions? In what ways can the government best avail itself on a continuing basis of the scientific community's special expertise in the evaluation and implementation of restrictions on the communication of scientific information?

The group should take special care to weigh the anticipated benefits of any restrictions against the costs of slowing scientific and technical progress, of which open scientific communication is an essential component. Therefore, in considering the above questions, the following concerns should also be addressed:

- In view of the unique educational mission of the Nation's academic institutions and their role in the advancement and dissemination of knowledge, are there types of government restrictions on the dissemination of information resulting from academic research that are less appropriate than others?
- What should be the government's policy with regard to information presented at open scientific meetings, and what procedures can be developed to ensure consistent implementation of that policy?

Administration

This study should be conducted by an interagency group chaired by the Office of Science and Technology Policy and including the Departments of State, Treasury, Defense, Justice, Commerce, Health and Human Services, Energy, the National Security Council staff, the Director of Central Intelligence, the Office of Management and Budget, the National Science Foundation, the National Aeronautics and Space Administration, and the General Services Administration. The scheduling and management of the Study is the responsibility of the Office of Science and Technology Policy.

A report for consideration by the National Security Council should be prepared no later than March 1, 1983.

APPENDIX D

National Telecommunications and Information Systems Security Policy No. 2: "National Policy on Protection of Sensitive, but Unclassified Information in Federal Government Telecommunications and Automated Information Systems"

SECTION I—POLICY

Federal Departments and Agencies shall ensure that telecommunications and automated information systems handling sensitive, but unclassified information will protect such information to the level of risk and the magnitude of loss or harm that could result from disclosure, loss, misuse, alteration, or destruction.

SECTION II—DEFINITION

Sensitive, but unclassified information is information the disclosure, loss, misuse, alteration, or destruction of which could adversely affect national security or other Federal Government interests. National security interests are those unclassified matters that relate to the national defense or the foreign relations of the U.S. Government. Other government interests are those related, but not limited to the wide range of government or government-derived economic, human, financial, industrial, agricultural, technological, and law enforcement information, as well as the privacy or confidentiality of personal or commercial proprietary information provided to the U.S. Government by its citizens.

SECTION III–APPLICABILITY

This policy applies to all Federal Executive Branch Departments and Agencies, and entities, including their contractors, which electronically transfer, store, process, or communicate sensitive, but unclassified information.

SECTION IV–RESPONSIBILITIES

This policy assigns to the heads of Federal Government Departments and Agencies the responsibility to determine what information is sensitive, but unclassified and to provide systems protection of such information which is electronically communicated, transferred, processed, or stored on telecommunications and automated information systems. The Director of Central Intelligence shall, in addition, be responsible for identifying sensitive, but unclassified information bearing on intelligence sources and methods and for establishing the system security handling procedures and the protection required for such information.

Federal Government Department and Agency heads shall:

a. Determine which of their department's or agency's information is sensitive, but unclassified and may warrant protection as sensitive during communications or processing via telecommunications or automated information systems. This determination should be based on the department's or agency's responsibilities, policies, and experience, and those requirements imposed by Federal statutes, as well as National Manager guidance on areas that potential adversaries have targeted;

b. Identify the systems which electronically process, store, transfer, or communicate sensitive, unclassified information requiring protection;

c. Determine, in coordination with the National Manager, as appropriate, the threat to and the vulnerability of those identified systems and;

d. Develop, fund and implement telecommunications and automated information security to the extent consistent with their mission responsibilities and in coordination with the National Manager, as appropriate, to satisfy their security or protection requirements.

e. Ensure implementation of telecommunications and automated information systems security consistent with the procedures and safeguards set forth in OMB Circular A-123 and A-130.

The National Manager shall, when requested, assist Federal Government Departments and Agencies to assess the threat to and vulnerability of targeted systems, to identify and document their telecommunications and automated information systems protection needs, and to develop the necessary security architecture.

BIBLIOGRAPHY

ARTICLES

Adler, Allan. "Unclassified Secrets." *Bulletin of the Atomic Scientists*, v. 41, March, 1985: 26–28.

Adler, Judith Lowitz. "The Impact of FOIA on Scientific Research Grantees." *Columbia Journal of Law and Social Problems*, v. 17, 1981: 1–44.

Aftergood, Steven. "The Perils of Government Secrecy." *Issues in Science and Technology*, v. 8, Summer, 1992: 81–88.

Alexander, Christine. "Preserving High Technology Secrets: National Security Controls on University Research and Teaching." *Law and Policy in International Business*, v. 15, 1983: 173–240.

Almond, Gabriel A. "Public Opinion and National Security Policy." *Public Opinion Quarterly*, v. 20, Summer, 1956: 371–378.

Anon. "AIAA Feels Chilling Effect Of DoD Censors." *Defense Week*, v. 5, May 7, 1984: 6.

______ . "Campus Unrest Affects Chemical Education." *Chemical and Engineering News*, v. 48, September 7, 1970: 38–40, 45–46.

Barnard, Richard. "Pentagon Mulls New Technology Secrets List." *Defense Week*, v. 4, October 17, 1983: 1, 5.

______ . "Pentagon Weighs New Strictures on Unclassified Papers." *Defense Week*, v. 4, October 24, 1983: 1, 10–11.

Becker, Jorg. "West-East Data Transfer: The German Connection." *Transnational Data and Communications Report*, v. 9, August, 1986: 11–14.

Benson, Sumner. "Overcoming Complacency." *Society*, v. 23, July-August, 1986: 12–15.

______ . "Technology Security: Political Lessons for Public Managers." *The Bureaucrat*, v. 15, Summer, 1986: 7–10.

Berkner, Lloyd V. "Secrecy and Scientific Progress." *Science*, v. 123, May 4, 1956: 783–786.

Berman, Jerry J. "National Security vs. Access to Computer Databases: A New Threat to Freedom of Information?" *Software Law Journal*, v. 2, Winter, 1987: 1–15.

Blados, Walter R. "Controlling Unclassified Scientific and Technical Information." *Information Management Review*, v. 2, Spring, 1987: 49–60.

Bock, P. G., and Morton Berkowitz. "The Emerging Field of National Security." *World Affairs*, v. 16, October, 1966: 122–136.

Bucy, J. Fred. "On Strategic Technology Transfer to the Soviet Union." *International Security*, v. 1, Spring, 1977: 25–43.

_____ . "Technology Transfer and East-West Trade: A Reappraisal." *International Security*, v. 5, Winter, 1980–1981: 132–151.

Carey, William D. "Classifying Science: . . . And a Scientist's Objection." *Aviation Week and Space Technology*, v. 116, February 8, 1982: 10–11, 82.

_____ . "Science and the National Security." *Science*, v. 214, November 6, 1981: 609.

_____ , and Frank Carlucci. "Scientific Exchanges and U.S. National Security." *Science*, v. 215, January 8, 1982: 139–141.

Carver, George A., Jr. "Intelligence in the Age of Glasnost." *Foreign Affairs*, v. 69, Summer, 1990: 147–166.

Chalk, Rosemary. "Commentary on the NAS Report." *Science, Technology, and Human Values*, v. 8, Winter, 1983: 21–24.

_____ . "Continuing Debate Over Science and Secrecy." *Bulletin of the Atomic Scientists*, v. 42, March, 1986: 14–16.

Cheh, Mary M. "Government Control of Private Ideas—Striking A Balance Between Scientific Freedom and National Security." *Jurimetrics Journal*, v. 23, Fall, 1982: 1–32.

_____ . "Judicial Supervision of Executive Secrecy: Rethinking Freedom of Expression for Government Employees and the Public Right of Access to Government Information." *Cornell Law Review*, v. 69, April, 1984: 690–734.

_____ . "The *Progressive* Case and the Atomic Energy Act: Waking to the Dangers of Government Information Controls." *George Washington Law Review*, v. 48, January, 1980: 163–210.

Childs, Harwood L. "Public Information and Opinion." *American Political Science Review*, v. 37, February, 1943: 56–68.

_____ , ed. "The Office of War Information." *Public Opinion Quarterly*, v. 7, Spring, 1943, whole issue.

Chimes, Lewis. "National Security and the First Amendment: The Proposed Use of Government Secrecy Agreements Under National Security Directive 84." *Columbia Journal of Law and Social Problems*, v. 19, 1985: 209–251.

Cohen, I. Bernard. "Science and the Growth of the American Republic." *Review of Politics*, v. 38, July, 1976: 359–398.

Coykendall, Robert W. "*In Camera* Inspection of National Security Files Under the Freedom of Information Act." *Kansas Law Review*, v. 26, Summer, 1978: 617–624.

David, Peter. "US Academics Jib at Pentagon Secrecy." *Nature*, 308, March 29, 1984: 389.

DeLauer, Richard D. "Scientific Communication and National Security." *Science*, v. 226, October 5, 1984: 9.

Denning, Peter J. "A Scientist's View of Government Control Over Scientific Publication." *IEEE Technology and Society Magazine*, v. v. 1, September, 1982: 17–18.

_____ . "Further Comments on National Security and Scientific Openness." *IEEE Technology and Society Magazine*, v. 1, September, 1982: 22–23.

_____ . "The Science of Computing: A New Paradigm for Science." *American Scientist*, v. 75, November-December, 1987: 572–573.

Dupree, A. Hunter. "National Security and the Post-War Science Establishment in the United States." *Nature*, v. 323, September 18, 1986: 213–216.

Eamon, William. "From the Secrets of Nature to Public Knowledge: The Origins of the Concept of Openness in Science." *Minerva*, v. 23, Autumn, 1985: 321–347.

_____ , and Francoise Paheau. "The Academia Segreta of Giolamo Ruscelli: A Sixteenth-Century Italian Scientific Society." *Isis*, v. 75, June, 1984: 327–342.

Edsall, John T. "Government and the Freedom of Science." *Science*, v. 121, April 29, 1955: 615–619.

______ . "Two Aspects of Scientific Responsibility." *Science*, v. 212, April 3, 1981: 11–14.

Ehlke, Richard C, and Harold C. Relyea. "The Reagan Administration Order on Security Classification: A Critical Assessment." *Federal Bar News and Journal*, v. 30, February, 1983: 91–97.

Ember, Lois R. "Secrecy in Science: A Contradiction in Terms?" *Chemical and Engineering News*, v. 60, April 5, 1982: 10–17.

______ . "U.S. Biotechnology Transfer Policy Being Studied." *Chemical and Engineering News*, v. 61, January 17, 1983: 25–26, 28.

Faberge, A. C. "Open Information and Secrecy in Research." *Perspectives in Biology and Medicine*, v. 25, Winter, 1982: 263–278.

Favre, David, and Matthew McKinnon. "The New Prometheus: Will Scientific Inquiry be Bound by the Chains of Government Regulation?" *Duquesne Law Review*, v. 19, Summer, 1981: 651–730.

Ferguson, James R. "Scientific and Technological Expression: A Problem in First Amendment Theory." *Harvard Civil Rights-Civil Liberties Law Review*, v. 16, Fall, 1981: 519–560.

______ . "Scientific Freedom, National Security, and the First Amendment." *Science*, v. 221, August 12, 1983: 620–624.

______ . "Scientific Inquiry and the First Amendment." *Cornell Law Review*, v. 64, April, 1979: 639–665.

Friedberg, Aaron L. "The Changing Relationship Between Economics and National Security." *Political Science Quarterly*, v. 106, Summer, 1991: 265–276.

Funk, Roger. "National Security Controls on the Dissemination of Privately Generated Scientific Information." *UCLA Law Review*, v. 30, December, 1982: 405–454.

Gerjuoy, Edward. "Embargo on Ideas: The Reagan Isolationism." *Bulletin of the Atomic Scientists*, v. 38, November, 1982: 31–37.

Glennon, Michael J. "The Gulf War and the Constitution." *Foreign Affairs*, v. 70, Spring, 1991: 84–101.

Goldberg, Steven. "The Constitutional Status of American Science." *University of Illinois Law Forum*, v. 1979, 1979: 1–33.

Goodman, Seymour E., Marjory S. Blumenthal, and Gary S. Geipel. "Export Control Reconsidered." *Issues in Science and Technology*, v. 6, Winter, 1989–1990: 58–62.

Gould, Stephen B. "Secrecy: Its Role in National Scientific and Technical Information Policy." *Library Trends*, v. 35, Summer, 1986: 61–81.

Gray, Paul E. "Technology Transfer at Issue: The Academic Viewpoint." *IEEE Spectrum*, v. 19, May, 1982: 64–68.

Green, Harold P. "The Boundaries of Scientific Freedom." *Newsletter on Science, Technology, and Human Values*, June, 1977: 17–24.

Greenberg, Joel. "Science's New Cold War." *Science News*, v. 123, April 2, 1983: 218–222.

Greenstein, Ruth. "National Security Controls on Scientific Information." *Jurimetrics Journal*, v. 23, Fall, 1982: 50–88.

Gregory, William H. "The Technology Transfer Mess." *Aviation Week and Space Technology*, v. 120, May 14, 1984: 13

Grove, J. W. "Science As Technology: Aspects of a Potent Myth." *Minerva*, v. 18, Summer, 1980: 293–312.

Hall, Marie Boas. "Oldenburg and the Art of Scientific Communication." *British Journal for the History of Science*, v. 2, December, 1965: 277–290.

Hart, Herbert L. "Are There Any Natural Rights?" *Philosophical Review*, v. 64, April, 1955: 175–191.

Herring, Jan P. "Defining the Basics." *Society*, v. 23, July-August, 1986: 16–19.

Holden, Constance. "Feds Defend Bubble Meddle." *Science*, v. 208, May 9, 1980: 577.

Horowitz, Irving Louis. "Revisiting the Dilemma of Rights and Obligations." *Society*, v. 23, July-August, 1986: 5–8.

Hull, David. "Openness and Secrecy in Science: Their Origins and Limitations." *Science, Technology, and Human Values*, v. 10, Spring, 1985: 4–13.

Hunter, Howard O. "The Constitutional Status of Academic Freedom in the United States." *Minerva*, v. 19, Winter, 1981: 519–568.

Inman, Bobby R. "A Response to Dr. Peter J. Denning's Article: 'A Scientist's View of Government Control Over Scientific Publication'." *IEEE Technology and Society Magazine*, v. 1, September, 1982: 21.

______. "Classifying Science: A Government Proposal . . . ." *Aviation Week and Space Technology*, v. 116, February 8, 1982: 10–11, 82.

______. "One View of National Security and Technical Information." *IEEE Technology and Society Magazine*, v. 1, September, 1982: 19–21.

______, and Daniel F. Burton, Jr. "Technology and Competitiveness: The New Policy Frontier." *Foreign Affairs*, v. 69, Spring, 1990: 116–134.

J., M. E. "Will Export Regulations Affect Academic Freedom?" *Physics Today*, v. 34, June, 1981: 55–57.

Kahn, David. "Cryptology Goes Public." *Foreign Affairs*, v. 58, Fall, 1979: 141-159.

______. "The Public's Secrets." *The Progressive*, v. 44, November, 1980: 27–31.

Kalivoda, Kenneth. "The Export Administration Act's Technical Data Regulations: Do They Violate the First Amendment?" *Georgia Journal of International and Comparative Law*, v. 11, Fall, 1981: 563–587.

Kalven, Jamie. "At War with the First Amendment." *Bulletin of the Atomic Scientists*, v. 45, January-February, 1989: 56–59.

Kevles, Daniel J. "George Ellery Hale, the First World War, and the Advancement of Science in America." *Isis*, v. 59, Winter, 1968: 427–437.

Knight, Jonathan. "Cryptographic Research and NSA." *Academe*, v. 67, December, 1981: 371-372.

Kolata, Gina Bari. "Attempts to Safeguard Technology Draw Fire." *Science*, v. 212, May 1, 1981: 523–524, 526.

______. "Export Control Threat Disrupts Meeting." *Science*, v. 217, September 24, 1982: 1233–1234.

______. "Security Controls Hurt Research, NAS Warns." *Science*, v. 218, October 15, 1982: 271-272.

______. "Technology Transfer: New Controls Urged." *Science*, v. 215, February 5, 1982: 635–636.

L., G. B. "Government Bars Soviets from AVS and OSA Meetings." *Physics Today*, v. 33, April, 1980: 81–83.

Ladenson, Robert F. "Scientific and Technical Information, National Security, and the First Amendment: A Jurisprudential Inquiry." *Public Affairs Quarterly*, v. 1, April, 1987: 1–20.

Langenberg, Donald N. "Secret Knowledge and Open Inquiry." *Society*, v. 23, July-August, 1986: 9–12.

Lepkowski, Wil. "Bromley at Midterm Is Confident, In Charge In Science Adviser Job." *Chemical and Engineering News*, v. 69, May 20, 1991: 21–23.

Levin, Ronald M. "In Camera Inspections Under the Freedom of Information Act." *University of Chicago Law Review*, v. 41, Spring, 1974: 557–581.

Long, Janice. "Scientific Freedom: Focus of National Security Controls Shifting." *Chemical and Engineering News*, v. 63, July 1, 1985: 7–11.

Lubrano, Linda. "National and International Politics in U.S.-U.S.S.R. Scientific Cooperation." *Social Studies in Science*, v. 11, November, 1981: 451–480.

Macdonald, Stuart. "Controlling the Flow of High-Technology Information from the

United States to the Soviet Union: A Labour of Sisyphus?" *Minerva*, v. 24, 1986: 39–73.

______. "US Export Controls and High-Tech Information." *Transnational Data and Communications Report*, v. 10, September, 1987: 18–23.

Madison, Christopher. "Administration Wins a Battle But Worries About the Export Control War." *National Journal*, v. 15, July 30, 1983: 1595–1597.

Magavero, Gerald. "The History and Background of American Copyright Law: An Overview." *International Journal of Law Librarians*, v. 6, July, 1978: 151–158.

Mally, Gerhard. "Technology Transfer Controls." *Jurimetrics Journal*, v. 23, Fall, 1982: 33–39.

Mann, Paul. "Defense Dept. Drafts Controls on Data." *Aviation Week and Space Technology*, v. 120, January 2, 1984: 17–18.

______. "Export Policy Triggers Dispute." *Aviation Week and Space Technology*, v. 119, December 19, 1983: 18–19.

______. "New Center to Oversee Export Licenses." *Aviation Week and Space Technology*, v. 119, September 19, 1983: 71–74.

Marshall, Eliot. "Do Seminars Leak Navy Secrets?" *Science*, v. 224, June 29, 1984: 1409.

______. "Export Law Affects Scientific Meetings." *Science*, v. 211, March 20, 1981: 1326.

______. "Medical Data Bank: A Security Risk?" *Science*, v. 216, May 21, 1982: 831.

McKelvain, Boyd J. "Determining Military Criticality." *Society*, v. 23, July- August, 1986: 19–21.

McMullin, Ernan. "Openness and Secrecy in Science: Some Notes on Early History." *Science, Technology, and Human Values*, v. 10, Spring, 1985: 14–23.

Medow, Jonathan C. "The First Amendment and the Secrecy State: *Snepp v. United States*." *University of Pennsylvania Law Review*, v. 130, April, 1982: 775–844.

Moran, Theodore H. "International Economics and National Security." *Foreign Affairs*, v. 69, Winter, 1990–1991: 74–90.

Mountain, Maurice J. "The Continuing Complexities of Technology Transfer." *Government Executive*, v. 11, January, 1979: 42–46, 48, 50, 52.

Murray, Alan. "High-Tech U.S. Exports Pose Dilemma for Policy-Makers." *Congressional Quarterly Weekly Report*, v. 41, March 26, 1983: 609–614.

Norman, Colin. "Academy Panel Blasts U.S. Export Controls." *Science*, v. 235, January 23, 1987: 424–425.

______. "DeLauer Questions DOD Censorship." *Science*, v. 224, May 4, 1984: 471.

______. "DOD Lists Critical Technologies." *Science*, v. 243, March 24, 1989: 1543.

______. "Illinois, Cornell Sign Supercomputer Contracts." *Science*, v. 229, August 9, 1985: 538.

______. "Memo Sets Policy for 'Star Wars' Publications." *Science*, v. 229, August 30, 1985: 843.

______. "Mixed Signals on Export Controls." *Science*, v. 226, December 14, 1984: 1295.

______. "Musical Chairs at OSTP." *Science*, v. 220, July 17, 1983: 1255.

______. "OSTP Seeks Advice on Export Controls." *Science*, v. 225, September 154, 1984: 1131.

______. "Pentagon Seeks to Build Bridges to Academe." *Science*, v. 228, April 19, 1985: 303, 305.

______. "Physics Institute Will Not Restrict Meetings." *Science*, v. 229, July 19, 1985: 253.

______. "Security Problems Plague Scientific Meeting." *Science*, v. 228, April 26, 1985: 471–472.

______. "Supercomputer Restrictions Pose Problems for NSF, Universities." *Science*, v. 229, July 12, 1985: 148.

______. "To Catch A Spy." *Science*, v. 222, November 25, 1983: 904–905.

______. "Universities Gag on Research Controls." *Science*, v. 224, April 13, 1984: 134.

______. "Universities Prevail on Secrecy." *Science*, v. 226, October 26, 1984: 418.

Note. "National Security and the Amended Freedom of Information Act." *Yale Law Journal*, v. 85, January, 1976: 401–422.

_____. "National Security and the Public's Right to Know: A New Role for the Courts Under the Freedom of Information Act." *University of Pennsylvania Law Review*, v. 123, June, 1975: 1438–1473.

Paine, Christopher. "Admiral Inman's Tidal Wave." *Bulletin of the Atomic Scientists*, v. 38, March, 1982: 3–6.

Park, Robert L. "Intimidation Leads to Self-Censorship in Science." *Bulletin of the Atomic Scientists*, v. 41, March, 1985: 22–25.

_____. "Restricting Information: A Dangerous Game." *Issues in Science and Technology*, v. 5, Fall, 1988: 62–67.

_____. "The Muzzling of American Science." *Academe*, v. 72, September-October, 1986: 19–23.

Paust, Jordan J. "Is the President Bound by the Supreme Law of the Land?—Foreign Affairs and National Security Reexamined." *Hastings Constitutional Law Review*, v. 9, Summer, 1982: 719–772.

Peterson, Ivars. "Controlling Technology Exports: Security vs. Knowledge." *Science News*, v. 121, March 20, 1982: 204, 206.

_____. "Silencing Science for Security." *Science News*, v. 121, January 16, 1982: 35.

_____, Janet Raloff, and Joel Greenberg. "'Remote Censoring:' DOD Blocks Symposium Papers." *Science News*, v. 122, September 4, 1982: 148–149.

Pike, John. "When Science Is Outlawed." *Inquiry*, v. 5, March 29, 1982: 21–25.

Price, Byron. "Governmental Censorship in War-time." *American Political Science Review*, v. 36, October, 1942: 837–850.

Quinn, James Brian, and Robert M. Cavanaugh. "Fundamental Research Can Be Planned." *Harvard Business Review*, v. 42, January-February, 1964: 111–124.

Rau, James Peter. "Government Secrecy Agreements and the First Amendment." *American University Law Review*, v. 28, Spring, 1979: 395–426.

Reingold, Nathan. "Science in the Civil War: The Permanent Commission of the Navy Department." *Isis*, v. 49, September, 1958: 307–318.

Relyea, Harold C. "Increased National Security Controls on Scientific Communication." *Government Information Quarterly*, v. 1, 1984: 173–207.

_____. "Information, Secrecy, and Atomic Energy." *New York University Review of Law and Social Change*, v. 10, 1980–1981: 265–286.

_____. "Scientific Communication and National Security: The Need for Balance." *National Forum*, v. 71, Fall, 1990: 34–37.

_____. "Shrouding the Endless Frontier—Scientific Communication and National Security: Considerations for a Policy Balance Sheet." *Government Information Quarterly*, v. 1, 1984: 1–14.

_____. "The Coming of Secret Law." *Government Information Quarterly*, v. 5, 1988: 97–116.

_____. "The Constitution and American Science." *Forum for Applied Research and Public Policy*, v. 2, Fall, 1987: 106–116.

Robins, Natalie. "The F.B.I.'s Invasion of Libraries." *The Nation*, v. 246, April 9, 1988: 481, 498–502.

Rosenbaum, Robert A., Morton J. Tenzer, Stephen H. Unger, William Van Alstyne, and Jonathan Knight. "Academic Freedom and the Classified Information System." *Science*, v. 219, January 21, 1983: 257–259.

Rosenzweig, Robert M. "Research as Intellectual Property: Influences Within the University." *Science, Technology, and Human Values*, v. 10, Spring, 1985: 41–48.

Rothschild, Lord. "Forty-Five Varieties of Research and Development." *Nature*, v. 239, October 13, 1972: 373–378.

Sagdeev, Roald Z. "Science and Perestroika: A Long Way to Go." *Issues in Science and Technology*, v. 4, Summer, 1988: 48–52.

Schmitt, Roland W. "Export Controls: Balancing Technological Innovation and National Security." *Issues in Science and Technology*, v. 1, Fall, 1984: 117–126.

Schultz, Charles L. "The Economic Content of National Security Policy." *Foreign Affairs*, v. 51, April, 1973: 522–540.

Shapley, Deborah. "Intelligence Agency Chief Seeks 'Dialogue' With Academics." *Science*, v. 202, October 27, 1978: 407–410.

Shattuck, John. "National Security a Decade After Watergate." *Democracy*, v. 3, Winter, 1983: 56–71.

______, and Muriel Morisey Spence. "The Dangers of Information Control." *Technology Review*, v. 91, April, 1988: 62–73.

Shinn, Allen M., Jr. "The First Amendment and the Export Laws: Free Speech on Scientific and Technical Matters." *George Washington Law Review*, v. 58, January, 1990: 368–403.

Smith, R. Jeffrey. "The Mysterious Chinese Luggage Incident." *Science*, 216, June 11, 1982: 1204–1205.

______. "White House Issues Secrecy Guideline." *Science*, v. 230, October 11, 1985: 152.

Smits, William H., Jr. "Significance of the Question of High-Technology Transfer to the Soviet Union and Soviet-Bloc States." *Technology In Society*, v. 8, 1986: 157–170.

______. "T^2: Transfer of High Technology from the United States to the Soviet Bloc: A Public Policy Issue." *International Journal of Public Administration*, v. 6, 1984: 245–277.

Sorensen, Theodore C. "Rethinking National Security." *Foreign Affairs*, v. 69, Summer, 1990: 1–18.

Souers, Sidney W. "Policy Formulation for National Security." *American Political Science Review*, v. 43, June, 1949: 534–543.

Sullivan, Robert E., and Nancy E. Bader. "The Application of Export Control Laws to Scientific Research at Universities." *Journal of College and University Law*, v. 9, 1982–83: 451–467.

Sun, Marjorie. "Strains in U.S.-Japan Exchanges." *Science*, v. 237, July 31, 1987: 476–478.

______. "U.S. Wants to Keep Eye on Biotech Exports." *Science*, v. 229, July 26, 1985: 368.

Symposium. "Limits of Scientific Inquiry." *Daedalus*, v. 107, Spring, 1978, whole issue.

Taylor, Maxwell D. "The Legitimate Claims of National Security." *Foreign Affairs*, v. 52, April, 1974: 577–594.

Teller, Edward. "Secrecy: The Road to Nowhere." *Technology Review*, v. 84, October, 1981: 12, 14.

Thomas, Lewis. "Scientific Frontiers and National Frontiers: A Look Ahead." *Foreign Affairs*, v. 62, Spring, 1984: 966–994.

Thompson, Elvia H. "Technology Security: Administrative Challenges for Public Managers." *The Bureaucrat*, v. 15, Summer, 1986: 11–15.

Tolchin. Susan J. "Halting the Erosion Of America's Critical Assets." *Issues in Science and Technology*, v. 9, Spring, 1993: 65–72.

Truxal, Carol. "Buying, Selling, and Trading Technology." *IEEE Spectrum*, v. 21, February, 1984: 58–65.

Ullman, Richard H. "Redefining Security." *International Security*, v. 8, Summer, 1983: 129–153.

Unger, Stephen H. "The Growing Threat of Government Secrecy." *Technology Review*, v. 84, February-March, 1982: 30–39, 84–85.

Verhoog, Henk. "The Responsibilities of Scientists." *Minerva*, v. 19, Winter, 1981: 582–604.

Wade, Nicholas. "Science Meetings Catch the U.S.-Soviet Chill." *Science*, v. 207, March 7, 1980: 1056, 1058.

Wallerstein, Mitchel B. "Nurturing a Dynamic System." *Society*, v. 23, July- August, 1986: 22–24.

_____. "Scientific Communication and National Security in 1984." *Science*, v. 224, May 4, 1984: 460–466.

_____, and Stephen B. Gould, eds. "A Delicate Balance: Scientific Communication vs. National Security." *Issues in Science and Technology*, v. 4, Fall, 1987: 42–55.

Wallich, Paul. "The Dilemma of Technology Transfer." *IEEE Spectrum*, v. 19, September, 1982: 66–70.

_____. "Technology Transfer at Issue: The Industry Viewpoint." *IEEE Spectrum*, v. 19, May, 1982: 69–73.

Walsh, John. "DOD Funds More Research in Universities." *Science*, v. 212, May 29, 1981: 1003–1004.

_____. "DOD Springs Surprise on Secrecy Rules." *Science*, v. 224, June 8, 1984: 1081.

_____. "New Pressure on Scientific Exchanges." *Science*, v. 215, February 5, 1982: 637–638.

_____. "Shunning Cryptocensorship." *Science*, v. 212, June 12, 1981: 1250.

_____. "Soviet-U.S. Exchanges Under Scrutiny." *Science*, v. 221, July 22, 1983: 346–348.

_____. "U.S.-Soviet Exchanges–Redefining Coexistence." *Science*, v. 233, August 22, 1986: 833–834.

Weart, Spencer R. "Scientists with a Secret." *Physics Today*, v. 29, February, 1976: 23–30.

Wolfers, Arnold. "'National Security' As An Ambiguous Symbol." *Political Science Quarterly*, v. 67,. December, 1952: 481–502.

Zilsel, Edgar. "The Genesis of the Concept of Scientific Progress." *Journal of the History of Ideas*, v. 6, June, 1945: 325–349.

BOOKS

Allison, Graham, and Gregory F. Treverton, eds. *Rethinking America's Security: Beyond Cold War to New World Order*. New York: W. W. Norton and Company, 1992.

Association of American Universities. *National Security Controls and University Research: Selected Readings*. Washington: Association of American Universities, 1987.

Barnett, Robert W. *Beyond War: Japan's Concept of Comprehensive National Security*. Washington: Pergamon-Brassey's, 1984.

Beard, Charles A. *The Idea of National Interest*. New York: Macmillan Company, 1934.

Becker, Carl L. *Freedom and Responsibility in the American Way of Life*. New York: Alfred A. Knopf, 1947.

Berkowitz, Morton R., and P. G. Bock, eds. *American National Security*. New York: Free Press, 1965.

Bertsch, Gary K., and John R. McIntyre, eds. *National Security and Technology Transfer*. Boulder: Westview Press, 1983.

Birch, Thomas. *History of the Royal Society of London*. New York: Johnson Reprint Corporation, 1968; originally published 1756.

Blechman, Barry M. *The Politics of National Security*. New York: Oxford University Press, 1990.

Blinken, Anthony J. *Ally Versus Ally*. New York: Praeger, 1987.

Bok, Sissela. *Secrets*. New York: Pantheon Books, 1982.

Brown, Anthony Cave, and Charles B. MacDonald, eds. *The Secret History of the Atomic Bomb*. New York: Dial Press/James Wade, 1977.

Bruce, Robert V. *Lincoln and the Tools of War*. Indianapolis: The Bobbs-Merrill Company, 1956.

———. *The Launching of Modern American Science, 1846–1876*. New York: Alfred A. Knopf, 1987.

Chafee, Zechariah, Jr. *Free Speech in the United States*. Cambridge: Harvard University Press, 1941.

Clem, Harold J., and Stanley L. Falk. *The Environment of National Security*. Washington: National Defense University Press, 1977.

Creel, George. *How We Advertised America*. New York: Harper and Brothers, 1920.

Cuff, Robert D. *The War Industries Board: Business-Government Relations During World War I*. Baltimore: The Johns Hopkins University Press, 1973.

Demac, Donna A. *Keeping America Uninformed*. New York: The Pilgrim Press, 1984.

deVolpi, A., G. E. Marsh, T. A. Postol, and G. S. Stanford. *Born Secret*. New York: Pergamon Press, 1981.

Dickson, David. *The New Politics of Science*. New York: Pantheon Books, 1984.

Douglas, Susan J. *Inventing American Broadcasting, 1899–1922*. Baltimore: The Johns Hopkins University Press, 1987.

Dupre, J. Stefan, and Sanford A. Lakoff, eds. *Science and the Nation*. Englewood Cliffs: Prentice-Hall, 1962.

Dupree, A. Hunter. *Science in the Federal Government: A History of Policies and Activities*. Baltimore: The Johns Hopkins University Press, 1986.

Emerson, Thomas I. *The System of Freedom of Expression*. New York: Random House, 1970.

England, Merton J. *A Patron for Pure Science*. Washington: National Science Foundation, 1982.

Finlay, David J., Ole R. Holsti, and Richard P. Fagen. *Enemies in Politics*. Chicago: Rand McNally, 1967.

Foerstel, Herbert N. *Surveillance in the Stacks: The FBI's Library Awareness Program*. New York: Greenwood Press, 1991.

Fowler, Dorothy Ganfield. *Unmailable: Congress and the Post Office*. Athens: The University of Georgia Press, 1977.

Friedrich, C. J. *Constitutional Reason of State: The Survival of the Constitutional Order*. Providence: Brown University Press, 1957.

Gellhorn, Walter. *Security, Loyalty, and Science*. Ithaca: Cornell University Press, 1950.

Gorbachev, Mikail. *Perestroika: New Thinking for Our Country and the World*. New York: Harper and Row, 1987.

Greenstein, Fred I., ed. *The Reagan Presidency: An Early Assessment*. Baltimore: The Johns Hopkins University Press, 1983.

Groves, Leslie R. *Now It Can Be Told: The Story of the Manhattan Project*. New York: Harper and Brothers, 1962.

Gruber, Carol S. *Mars and Minerva: World War I and the Uses of Higher Learning in America*. Baton Rouge: Louisiana State University Press, 1975.

Gwertzman, Bernard, and Michael T. Kaufman, eds. *The Collapse of Communism*. New York: Times Books, 1990.

Hahn, Roger. *The Anatomy of a Scientific Institution*. Berkeley: University of California Press, 1971.

Hawke, David Freeman. *Nuts and Bolts of the Past*. New York: Harper and Row, 1988.

Hewlett, Richard G., and Oscar E. Anderson, Jr. *A History of the United States Atomic Energy Commission: The New World, 1939/1946*. University Park: Pennsylvania State University Press, 1962.

Hewlett, Richard G., and Jack M. Holl. *Atoms For Peace and War, 1953–1961*. Berkeley: University of California Press, 1989.

Hofstadter, Richard, and Walter P. Metzger. *The Development of Academic Freedom in the United States*. New York: Columbia University Press, 1955.

Holton, Gerald, and Robert S. Morison, eds. *The Limits of Scientific Inquiry*. New York: W. W. Norton and Company, 1979.

Hughes, Thomas P. *American Genesis*. New York: Viking Penguin, 1989.

Hull, Elizabeth. *Taking Liberties*. New York: Praeger, 1990.

Huntington, Samuel P. *The Soldier and the State: The Theory and Politics of Civil-Military Relations*. Cambridge: The Belknap Press of Harvard University Press, 1957.

Kaplan, Abraham. *The Conduct of Inquiry*. San Francisco: Chandler Publishing Company, 1964.

Kemeny, John G. *A Philosopher Looks At Science*. Princeton: D. Van Nostrand Company, 1959.

Koh, Harold Hongju. *The National Security Constitution: Sharing Power After the Iran-Contra Affair*. New Haven: Yale University Press, 1990.

Kuhn, Thomas S. *The Structure of Scientific Revolutions*. Second edition. Chicago: University of Chicago Press, 1970.

Landau, Saul. *The Dangerous Doctrine: National Security and U. S. Foreign Policy*. Boulder: Westview Press, 1988.

Langfitt, Thomas W., Sheldon Hackney, Alfred P. Fishman, and Albert V. Glowasky, eds. *Partners In The Research Enterprise*. Philadelphia: University of Pennsylvania Press, 1983.

Lasswell, Harold D. *National Security and Individual Freedom*. New York: McGraw-Hill Book Company, 1950.

Lippmann, Walter. *U. S. Foreign Policy: Shield of the Republic*. Boston: Little, Brown and Company, 1943.

Lloyd, G. E. R. *Magic, Reason, and Experience*. Cambridge: Cambridge University Press, 1979.

Long, William J. *U. S. Export Control Policy*. New York: Columbia University Press, 1989.

Macdonald, Stuart. *Technology and the Tyranny of Export Controls*. Houndmills: The Macmillan Press, 1990.

McCamy, James L. *Government Publicity: Its Practice in Federal Administration*. Chicago: University of Chicago Press, 1939.

Meinecke, Friedrich. *Machiavellism: The Doctrine of Raison D'Etat and Its Place in Modern History*. New York: Frederick A. Praeger, 1965.

Melvern, Linda, David Hebditch, and Nick Anning. *Techno-Bandits*. Boston: Houghton Mifflin Company, 1984.

Mercer, J. Edward. *Alchemy: Its Science and Romance*. London: Macmillan Company, 1921.

Merton, Robert K., ed. *The Sociology of Science*. Chicago: University of Chicago Press, 1973.

Metcalfe, Robyn Shotwell. *The New Wizard War*. Redmond: Tempus Books, 1988.

Mock, James R. *Censorship 1917*. Princeton: Princeton University Press, 1941.

______ , and Cedric Larson. *Words that Won the War*. Princeton: Princeton University Press, 1939.

Murphy, Paul L. *World War I and the Origin of Civil Liberties in the United States*. New York: W. W. Norton and Company, 1979.

Nagel, Ernest. *The Structure of Science*. New York: Harcourt, Brace, and World, 1961.

Nelkin, Dorothy. *Science As Intellectual Property*. New York: Macmillian Publishing Company, 1984.

______ . *The University and Military Research: Moral Politics at MIT*. Ithaca: Cornell University Press, 1972.

Nieburg, H. L. *In the Name of Science*. Revised edition. Chicago: Quadrangle Books, 1970.

Ornstein, Martha. *The Role of Scientific Societies in the Seventeenth Century*. Chicago: University of Chicago Press, 1913.

Paine, Lauran. *Silicon Spies*. New York: St. Martin's Press, 1986.

Patterson, Lyman R. *Copyright in Historical Perspective.* Nashville: Vanderbilt University Press, 1968.

Pell, Eve. *The Big Chill.* Boston: Beacon Press, 1984.

Price, Don K. *Government and Science.* New York: Oxford University Press, 1962.

______. *The Scientific Estate.* Cambridge: The Belknap Press of Harvard University Press, 1965.

Read, John. *Prelude to Chemistry; And An Outline of Alchemy.* New York: Macmillan Company, 1937.

Reingold, Nathan. *Science, American Style.* New Brunswick: Rutgers University Press, 1991.

Relyea, Harold C., ed. *Striking A Balance: National Security and Scientific Freedom—First Discussions.* Washington: American Association for the Advancement of Science, 1985.

______. *The Presidency and the People's Right to Know.* New York: Center for the Study of the Presidency, 1981.

Roberts, Philip A. *Technology Transfer: A Policy Model.* Washington: National Defense University Press, 1988.

Roland, Alex. *Model Research: The National Advisory Committee for Aeronautics, 1915–1958.* Washington: National Aeronautics and Space Administration, 1985.

Rossi, Paolo. *Philosophy, Technology, and the Arts in the Early Modern Era.* New York: Harper and Row, 1970.

Rostow, W. W. *How It All Began: Origins of the Modern Economy.* New York: McGraw-Hill Book Company, 1975.

Scheiber, Harry. *The Wilson Administration and Civil Liberties, 1917–1921.* Ithaca: Cornell University Press, 1960.

Schweizer, Peter. *Friendly Spies.* New York: Atlantic Monthly Press, 1993.

Scott, Lloyd N. *Naval Consulting Board of the United States.* Washington: U. S. Govt. Print. Off., 1920.

Smith, Hedrick. *The Power Game: How Washington Works.* New York: Random House, 1988.

Smith, Merritt Roe, ed. *Military Enterprise and Technological Change: Perspectives on the American Experience.* Cambridge: The M.I.T. Press, 1985.

Stevenson, Russell B., Jr. *Corporations and Information: Secrecy, Access, and Disclosure.* Baltimore: The Johns Hopkins University Press, 1980.

Stewart, Irwin. *Organizing Scientific Research for War: The Administrative History of the Office of Scientific Research and Development.* Boston: Little, Brown and Company, 1948.

Strasser, Gabor, and Eugene M. Simons, eds. *Science and Technology Policies.* Cambridge: Ballinger Publishing Company, 1973.

Summers, Robert E., comp. *Wartime Censorship of Press and Radio.* New York: H. W. Wilson Company, 1942.

______, ed. *Federal Information Controls in Peacetime.* New York: H. W. Wilson Company, 1949.

Taylor, Sherwood. *The Alchemists.* Woodstock: Beckman Publishers, 1977.

Tolchin, Martin, and Susan J. Tolchin. *Selling Our Security: The Erosion of America's Assets.* New York: Alfred A. Knopf, 1992.

Vaughn, Stephen L. *Holding Fast the Inner Lines: Democracy, Nationalism, and the Committee on Public Information.* Chapel Hill: The University of North Carolina Press, 1980.

Weil, Vivian, and John W. Snapper, eds. *Owning Scientific and Technical Information: Value and Ethical Issues.* New Brunswick: Rutgers University Press, 1989.

White, Stephen. *Gorbachev in Power.* Cambridge: Cambridge University Press, 1990.

Whitfield, Stephen J. *The Culture of the Cold War.* Baltimore: The Johns Hopkins University Press, 1991.

Wiggins, James Russell. *Freedom or Secrecy.* Revised edition. New York: Oxford University Press, 1964.

Willoughby, William Franklin. *Government Organization in War Time and After*. New York: D. Appleton and Company, 1919.

Yergen, Daniel R. *Shattered Peace: The Origins of the Cold War and the National Security State*. Boston: Houghton Mifflin Company, 1978.

Yerkes, Robert M., ed. *The New World of Science: Its Development During the War*. New York: The Century Company, 1920.

CONGRESSIONAL HEARINGS

U. S. Congress. House. Committee on Armed Services. *Full Committee Consideration of H.R. 4681*. Hearing, 98th Congress, 2d Session. Washington: U. S. Govt. Print. Off., 1984.

______. *Full Committee Consideration of H.R. 3231 to Amend the Authorities Contained in the Export Administration Act of 1979, and for Other Purposes*. Hearing, 98th Congress, 1st Session. Washington: U. S. Govt. Print. Off., 1984.

______. *Hearings on H.R. 4681*. Hearing, 98th Congress,m 2d Session. Washington: U. S. Govt. Print. Off., 1985.

______. *Technology Transfer*. Hearings, 98th Congress, 1st Session. Washington: U. S. Govt. Print. Off., 1984.

______. Committee on Banking, Finance, and Urban Affairs. *Proposed Joint Development of the FSX Fighter With Japan*. Hearings, 101st Congress, 1st Session. Washington: U. S. Govt. Print. Off., 1989.

______. Committee on Foreign Affairs. *Extension and Revision of the Export Administration Act of 1979*. Hearings and mark-up, 98th Congress, 1st Session. Washington: U. S. Govt. Print. Off., 1985.

______. *National Security Export Controls: Report by the National Academy of Sciences*. Hearing, 100th Congress, 1st Session. Washington: U. S. Govt. Print. Off., 1988.

______. *Technology Exports: Department of Defense Organization and Performance*. Hearing, 96th Congress, 1st Session. Washington: U. S. Govt. Print. Off., 1980.

______. *The Reauthorization of the Export Administration Act*. Hearings and mark-up, 101st Congress, 2d Session. Washington: U. S. Govt. Print. Off., 1990.

______. *The Reauthorization of the Export Administration Act*. Hearings and mark-up, 102d Congress, lst Session. Washington: U. S. Govt. Print. Off., 1992.

______. Committee on Government Operations. *Availability of Information from Federal Departments and Agencies*. Hearings, 84th-86th Congresses. Washington: U. S. Govt. Print. Off., 1956–1959.

______. *Computer Security Act of 1987*. Hearings, 100th Congress, 1st Session. Washington: U. S. Govt. Print. Off., 1987.

______. *Congress and the Administration's Secrecy Pledges*. Hearing, 100th Congress, 2d Session. Washington: U. S. Govt. Print. Off., 1988.

______. *Favoritism and Bias Within the Defense Science Board and Other Military Advisory Panels*. Hearing, 98th Congress, 1st Session. Washington: U. S. Govt. Print. Off., 1984.

______. *Military and Civilian Control of Computer Security Issues*. Hearing, 101st Congress, 1st Session. Washington: U. S. Govt. Print. Off., 1989.

______. *National Security Decision Directive 84 and the Proposed Department of Defense Directive on Polygraph Use*. Hearing, 98th Congress, 1st Session. Washington: U. S. Govt. Print. Off., 1984.

______. *The Government's Classification of Private Ideas*. Hearings, 96th Congress, 2d Session. Washington: U. S. Govt. Print. Off., 1981.

______. Committee on the Judiciary. *FBI Counterintelligence Visits To Libraries*. Hearings, 100th Congress, 2d Session. Washington: U. S. Govt. Print. Off., 1989.

______ . *Federal Polygraph Limitation and Anti-Censorship Act of 1984*. Hearing, 98th Congress, 2d Session. Washington: U. S. Govt. Print. Off., 1986.

______ . *1984: Civil Liberties and the National Security State*. Hearings, 98th Congress, 1st and 2d Sessions. Washington: U. S. Govt. Print. Off., 1984.

______ . *Prepublication Review and Secrecy Requirements Imposed Upon Federal Employees*. Hearings, 96th Congress, 2d Session. Washington: U. S. Govt. Print. Off., 1980.

______ . *Presidential Directive on the Use of Polygraphs and Prepublication Review*. Hearings, 98th Congress, 1st and 2d Sessions. Washington: U. S. Govt. Print. Off., 1985.

______ . Committee on Post Office and Civil Service. *Classified Information Nondisclosure Agreements*. Hearing, 100th Congress, 1st Session. Washington: U. S. Govt. Print. Off., 1988.

______ . *Federal Polygraph Limitation and Anti-Censorship Act of 1984*. Hearing, 98th Congress, 1st Session. Washington: U. S. Govt. Print. Off., 1984.

______and Committee on the Judiciary. *Protection of Classified Information*. Joint hearing, 99th Congress, 1st Session. Washington: U. S. Govt. Print. Off., 1987.

______ . Committee on Science and Technology. *Commercialization of Academic Biomedical Research*. Hearings, 97th Congress, 1st Session. Washington: U. S. Govt. Print. Off., 1981.

______ . *Impact of National Security Considerations on Science and Technology*. Hearing, 97th Congress, 2d Session. Washington: U. S. Govt. Print. Off., 1982.

______ . *Joint Industry/University Cooperation With Federally Supported Research Facilities*. Hearing, 98th Congress, 1st Session. Washington: U. S. Govt. Print. Off., 1983.

______ . *Scientific Communications and National Security*. Hearing, 98th Congress, 2d Session. Washington: U. S. Govt. Print. Off., 1984.

______ . *The Availability of Japanese Scientific and Technical Information in the United States*. Hearings, 98th Congress, 2d Session. Washington: U. S. Govt. Print. Off., 1984.

______ . Committee on Science, Space, and Technology. *FSX*. Hearings, 101st Congress, 1st Session. Washington: U. S. Govt. Print. Off., 1989.

______ . *Implementation of the Computer Security Act*. Hearing, 101st Congress, 1st Session. Washington: U. S. Govt. Print. Off., 1989.

______ . *Implementation of the Computer Security Act (Public Law 100–235)*. Hearing, 101st Congress, 2d Session. Washington: U. S. Govt. Print. Off., 1990.

______ . *National Academy of Sciences Report on International Technology Transfer*. Hearings, 100th Congress, 1st Session. Washington: U. S. Govt. Print. Off., 1987.

______ . *Technology Transfer*. Hearings, 99th Congress, 1st Session. Washington: U. S. Govt. Print. Off., 1985.

______ . Permanent Select Committee on Intelligence. *Prepublication Review and Secrecy Agreements*. Hearings, 96th Congress, 2d Session. Washington: U. S. Govt. Print. Off., 1980.

______ . Joint Economic Committee. *American Economic Power: Redefining National Security for the 1990's*. Hearings, 101st Congress, 1st and 2d Sessions. Washington: U. S. Govt. Print. Off., 1991.

______ . Senate. Committee on Banking, Housing, and Urban Affairs. *East-West Trade and Technology Transfer*. Hearing, 97th Congress, 2d Session. Washington: U. S. Govt. Print. Off., 1982.

______ . *Export Administration Act*. Hearing, 98th Congress, 1st Session. Washington: U. S. Govt. Print. Off., 1983.

______ . *Reauthorization of the Export Administration Act*. Hearings, 98th Congress, 1st Session. Washington: U. S. Govt. Print. Off., 1983.

______ . *Reauthorization of the Export Administration Act*. Hearings, 101st Congress, 2d Session. Washington: U. S. Govt. Print. Off., 1990.

______ . Committee on Finance. *National Research Council Report on International Competition in Advanced Technology*. Hearing, 98th Congress, 1st Session. Washington: U. S. Govt. Print. Of.., 1983.

_____ . Committee on Governmental Affairs. *National Security Decision Directive 84*. Hearing, 98th Congress, 1st Session. Washington: U. S. Govt. Print. Off., 1984.

_____ . *Transfer of Technology*. Hearings, 98th Congress, 2d Session. Washington: U. S. Govt. Print. Off., 1984.

_____ . *Transfer of United States High Technology to the Soviet Union and Soviet Bloc Nations*. Hearings, 97th Congress, 2d Session. Washington: U. S. Govt. Print. Off., 1982.

_____ . Committee on the Judiciary. *Information and Competitiveness*. Hearing, 100th Congress, 2d Session. Washington: U. S. Govt. Print. Off., 1989.

_____ . Committee on Military Affairs. *Department of Armed Forces, Department of Military Security*. Hearings, 79th Congress, 1st Session. Washington: U. S. Govt. Print. Off., 1945.

_____ . Special Committee on the Termination of the National Emergency. *National Emergency*. Hearings,. 93rd Congress, 1st Session. Washington: U. S. Govt. Print. Off., 1973.

REPORTS

Advanced Technology Systems, Inc. *Control of Unclassified Technology with Military Application*. April 15, 1983. Vienna, Virginia.

American Association for the Advancement of Science. "National Security and Scientific Communication: A Summary of Responses Received in Reply to a Letter from the AAAS Committee on Scientific Freedom and Responsibility." Washington: American Association for the Advancement of Science, June, 1982.

_____ . "National Security and Scientific Communication: Professional Society Chronology." Washington: American Association for the Advancement of Science, April, 1984.

Bushman, Cynthia, and Kathryn M. Porter. "Secrecy in Science." *Freedom of Information Report No. 429*. November, 1980. School of Journalism, University of Missouri, Columbia, Missouri.

Business-Higher Education Forum. *Export Controls: The Need to Balance National Objectives*. Washington: Business-Higher Education Forum, 1986.

Committee on Federal Legislation. "National Security An The First Amendment–NSDD 84." *Record of the Association of the Bar of the City of New York*, v. 39, April, 1984: 201–226.

Committee on Loyalty in Relation to Government Support of Unclassified Research. "Loyalty and Research." *Science*, v. 123, April 20, 1956: 660–662.

Committee on Science in the Promotion of Human Welfare. "Science and Human Welfare." *Science*, v. 132, July 8, 1960: 68–73.

_____ . "Secrecy and Dissemination in Science and Technology." *Science*, v. 163, February 21, 1969: 787–790.

_____ . "The Integrity of Science." *American Scientist*, v. 53, June, 1965: 174- 198.

CSIS Science and Technology Committee. Panel on Technology Transfer. *Securing Technological Advantage: Balancing Export Controls and Innovation*. Washington: Center for Strategic and International Studies, 1985.

Davis, Elmer, and Byron Price. *War Information and Censorship*. Washington: American Council on Public Affairs, 1943.

Director of Central Intelligence. *Soviet Acquisition of Western Technology*. April, 1982. Washington, D. C..

_____ . *Soviet Acquisition of Militarily Significant Western Technology: An Update*. September, 1985. Washington, D. C.

Edsall, John T. *Scientific Freedom and Responsibility.* Washington: American Association for the Advancement of Science, 1975.

Friedland, M. L. *National Security: The Legal Dimensions.* Hull: Supply and Services Canada, 1979.

Gustafson, Thane. *Selling the Russians the Rope? Soviet Technology Policy and U. S. Export Controls.* R-2649-ARPA. Santa Monica: The Rand Corporation, April, 1981.

Holwill, Richard N., ed. *Agenda '83: A Mandate for Leadership Report.* Washington: The Heritage Foundation, 1983.

Interim Committee on the Social Aspects of Science. "Social Aspects of Science." *Science,* v. 125, January 25, 1957: 143–147.

Massachusetts Institute of Technology. Committee on the Changing Nature of Information. *Interim Report of the Committee on the Changing Nature of Information.* March 9, 1983. Cambridge, Massachusetts.

Media Outreach Program. *National Security and Scientific Inquiry.* May 3, 1985. Washington, D. C.

Mellon Institute. Carnegie-Mellon University. *Proceedings of the Conference on the Political Freedom of Foreign Nationals on U. S. Campuses.* March 5, 1983. Pittsburgh, Pennsylvania.

National Academy of Sciences. *Balancing the National Interest.* Washington: National Academy Press, 1987.

______ . *Balancing the National Interest: Working Papers.* Washington: National Academy Press, 1987.

______ . *Computers At Risk.* Washington: National Academy Press, 1991.

______ . *Finding Common Ground: U. S. Export Controls in a Changed Global Environment.* Washington: National Academy Press, 1991.

______ . *International Competition in Advanced Technology: Decisions for America.* Washington: National Academy Press, 1983.

______ . *Report of the National Academy of Sciences for the Year 1913.* Washington: U. S. Govt. Print. Off., 1914.

______ . *Report of the National Academy of Sciences for the Year 1916.* Washington: U. S. Govt. Print. Off., 1917.

______ . *Report of the National Academy of Sciences for the Year 1917.* Washington: U. S. Govt. Print. Off., 1918.

______ . *Report of the National Academy of Sciences for the Year 1918.* Washington: U. S. Govt. Print. Off., 1919.

______ . *Report of the National Academy of Sciences for the Year 1919.* Washington: U. S. Govt. Print. Off., 1920.

______ . *Scientific Communication and National Security.* Washington: National Academy Press, 1982.

National Research Council. *Global Trends in Computer Technology and Their Impact on Export Control.* Washington: National Academy Press, 1988.

Naval Consulting Board of the United States. *Problems of Aeroplane Improvement.* Bulletin No. 3. New York: Naval Consulting Board, 1918.

______ . *The Enemy Submarine.* Bulletin No. 2. New York: Naval Consulting Board, 1918.

______ . *The Submarine and Kindred Problems.* Bulletin No. 1. New York: Naval Consulting Board, 1917.

President's Export Council. Export Administration Subcommittee. *Recommendations on Amending the Export Administration Act of 1979.* March, 1983. Washington, D. C.

President's Scientific Research Board. *Science and Public Policy.* Washington: U. S. Govt. Print. Off., 1947.

Public Cryptography Study Group. "Report of the Public Cryptography Study Group." *Academe,* v. 67, December, 1981: 372–382.

Rabkin, Yakov M. *Science Between the Superpowers.* A Twentieth Century Fund Paper. New York: Priority Press, 1988.

Slamecka, Vladimir. *Trade-Offs in International Flows of Scientific and Technical Information.* Atlanta: Georgia Institute of Technology, School of Information and Computer Science, September, 1980.

Special Committee on Civil Liberties for Scientists. "Civil Liberties of Scientists." *Science,* v. 110, August 19, 1949: 177–179.

U. S. Commission on Organization of the Executive Branch of the Government. *Task Force Report on National Security Organization.* Washington: U. S. Govt. Print. Off., 1949.

______ . *The National Security Organization.* Washington: U. S. Govt. Print. Off., 1949.

U. S. Committee on Public Information. *Complete Report of the Chairman of the Committee on Public Information 1917: 1918: 1919.* Washington: U. S. Govt. Print. Off., 1920.

U. S. Congress. House. Committee on Armed Services. *Federal Polygraph Limitation and Anti-Censorship Act of 1984.* H. Rept. 98-961 Part 2, 98th Congress, 2d Session. Washington: U. S. Govt. Print. Off., 1984.

______ . Committee on Foreign Affairs. *Export Administration Act Amendments of 1983.* H. Rept. 98-257 Part 1, 98th Congress, 1st Session. Washington: U. S. Govt. Print. Off., 1983.

______ . *Export Facilitation Act of 1990.* H. Rept. 101-482, 101st Congress, 2d Session. Washington: U. S. Govt. Print. Off., 1990.

______ . *The War Powers Resolution.* Committee print, 97th Congress, 2d Session. Washington: U. S. Govt. Print. Off., 1982.

______ . Committee on Government Operations. *Availability of Information from Federal Departments and Agencies.* H. Rept. 2947, 84th Congress, 2d Session. Washington: U. S. Govt. Print. Off., 1956.

______ . *Availability of Information from Federal Departments and Agencies.* H. Rept. 2578, 85th Congress, 2d Session. Washington: U. S. Govt. Print. Off., 1958.

______ . *Defense Science Boards: A Question of Integrity.* H. Rept. 98-580, 98th Congress, 1st Session. Washington: U. S. Govt. Print. Off., 1983.

______ . *Executive Classification of Information—Security Classification Problems Involving Exemption (b)(1) of the Freedom of Information Act (5 U.S.C. 552).* H. Rept. 93-221, 93rd Congress, 1st Session. Washington: U. S. Govt. Print. Off., 1973.

______ . *Security Classification Policy and Executive Order 12356.* H. Rept. 97-731, 97th Congress, 2d Session. Washington: U. S. Govt. Print. Off., 1982.

______ . *The Administration's Initiatives to Expand Polygraph Use and Impose Lifelong Censorship on Thousands of Government Employees.* H. Rept. 98-578, 98th Congress, 1st Session. Washington: U. S. Govt. Print. Off., 1984.

______ . *The Government's Classification of Private Ideas.* H. Rept. 96-1540, 96th Congress, 2d Session. Washington: U. S. Govt. Print. Off., 1980.

______ . Committee on Post Office and Civil Service. *Federal Polygraph Limitation and Anti-Censorship Act of 1984.* H. Rept. 98-961 Part 1, 98th Congress, 1st Session. Washington: U. S. Govt. Print. Off., 1984.

______ . Committee on Science and Technology. *A History of Science Policy in the United States, 1940–1985* by Jeffrey K. Stine. Committee print, 99th Congress, 2d Session. Washington: U. S. Govt. Print. Off., 1986.

______ . *Science Policy: A Working Glossary.* Fourth Edition. Prepared by the Science Policy Research Division, Congressional Research Service, Library of Congress. Committee print, 95th Congress, 2d Session. Washington: U. S. Govt. Print. Off., 1978.

______ . *Science Support by the Department of Defense.* Committee print, 99th Congress, 2d session. Washington: U. S. Govt. Print. Off., 1987.

______ . *The Availability of Japanese Scientific and Technical Information in the United*

States. Committee print, 98th Congress, 2d Session. Washington: U. S. Govt. Print. Off., 1984.

______. Committee on Science, Space, and Technology. *Export Controls, Competitiveness and International Cooperation: A Critical Review.* Committee print, 101st Congress, 1st Session. Washington: U. S. Govt. Print. Off., 1989.

______. Permanent Select Committee on Intelligence. *Security Clearance Procedures in the Intelligence Agencies: Staff Report.* Committee print, 96th Congress, 1st Session. Washington: U. S. Govt. Print. Off., 1979.

______. Select Committee to Investigate Covert Arms Transactions with Iran. [and] Senate. Select Committee on Secret Military Assistance to Iran and the Nicaraguan Opposition. *Report of the Congressional Committees Investigating the Iran-Contra Affair.* H. Rept. 100–433 and S. Rept. 100–216, 100th Congress, 1st Session. Washington: U. S. Govt. Print. Off., 1987.

______. Senate. Committee on Banking, Housing, and Urban Affairs. *The Export Administration Act Amendments of 1983.* S. Rept. 98–170, 98th Congress, 1st Session. Washington: U. S. Govt. Print,. Off., 1983.

______. Committee on Foreign Relations. *The U. S. Government and the Vietnam War.* Committee print, 98th Congress, 2d Session [and] 100th Congress, 2d Session. Washington: U. S. Govt. Print. Off., 1984, 1985, 1988.

______. Committee on Governmental Affairs. *Transfer of United States High Technology to the Soviet Union and Soviet Bloc Nations.* S. Rept. 97–664, 97th Congress, 2d Session. Washington: U. S. Govt. Print. Off., 1982.

______. Committee on Government Operations [and] Special Committee on National Emergencies and Delegated Emergency Powers. *The National Emergencies Act (Public Law 94–412). Source Book: Legislative History, Texts, and Other Documents.* Committee print, 94th Congress, 2d Session. Washington: U. S. Govt. Print. Off., 1976.

______. Committee on Naval Affairs. *Unification of the War and Navy Departments and Postwar Organization for National Security.* Committee print, 79th Congress, 1st Session. Washington: U. S. Govt. Print. Off., 1945.

______. Special Committee on National Emergencies and Delegated Emergency Powers. *National Emergencies and Delegated Emergency Powers.* S. Rept. 94–922, 94th Congress, 2d Session. Washington: U. S. Govt. Print. Off., 1976.

______. Special Committee on the Termination of the National Emergency. *Emergency Powers Statutes.* S. Rept. 93–549, 93rd Congress, 1st Session. Washington: U. S. Govt. Print. Off., 1973.

U. S. Department of Defense. Office of the Director of Defense Research and Engineering. *An Analysis of Export Control of U.S. Technology—A DOD Perspective: A Report of the Defense Science Board Task Force on Export of U.S. Technology.* February 4, 1976. Washington, D. C.

______. *Report of the Defense Science Board Task Force on Secrecy.* July 1, 1970. Washington, D. C.

______. Office of the Secretary of Defense. *The Technology Security Program: A Report to the 99th Congress.* February, 1985. Washington, D. C.

______. *The Technology Transfer Control Program: A Report to the 98th Congress.* February, 1983. Washington, D. C.

______. *The Technology Transfer Control Program: A Report to the 98th Congress.* February, 1984. Washington, D. C.

______. Office of the Under Secretary of Defense for Policy. *Assessing the Effect of Technology Transfer on U.S./Western Security: A Defense Perspective.* February, 1985. Washington, D. C.

______. Office of the Under Secretary of Defense for Research and Engineering. *Report of the Defense Science Board Task Force on University Responsiveness to National Security Requirements.* January, 1982. Washington, D. C.

U. S. Department of State. *Science and Foreign Relations.* Washington: Department of State, 1950.

U. S. General Accounting Office. *Export Control Regulation Could Be Reduced Without Affecting National Security.* GAO/IS-82-14. Washington: U. S. Govt. Print. Off., 1982.

______. *Federal Agency Use of Nondisclosure Agreements.* GAO/NSIAD-91- 106FS. Washington: U. S. Govt. Print. Off., 1991.

______. *Information Security: Update of Data on Employees Affected by Federal Security Programs.* GAO/NSIAD-89-56FS. Washington: U. S. Govt. Print. Off., 1989.

______. *Information and Personnel Security: Data on Employees Affected by Federal Security Programs.* GAO/NSIAD-86-189FS. Washington: U. S. Govt. Print. Off., 1986.

U. S. National Science Foundation. Mathematical and Computer Sciences Advisory Committee. *The Role of the NSF in Supporting Cryptological Research.* July 13, 1981. Washington, D. C.

______. National Science Board. *Report of the NSB Committee on Openness of Scientific Communication.* December 2, 1988. Washington, D. C.

______. *University-Industry Research Relationships: Myths, Realities, and Potentials.* Fourteenth Annual Report of the National Science Board to the President and the Congress. Washington: National Science Foundation, 1982.

______. *University-Industry Research Relationships: Selected Studies.* Washington: National Science Foundation, 1983.

U. S. Office of Scientific Research and Development. *Science—The Endless Frontier* by Vannevar Bush. Washington: U. S. National Science Foundation, 1960; originally published 1945.

U. S. Office of Science and Technology Policy. *Report of the Working Group on Basic Research in the Department of Defense.* June 22, 1978. Washington, D. C.

______. *U. S. Technology Policy.* September 26, 1990. Washington, D. C.

U. S. Office of Technology Assessment. *Adjusting to a New Security Environment: The Defense Technology and Industrial Base Challenge.* Washington: U. S. Govt. Print. Off., 1991.

______. *Arming Our Allies: Cooperation and Competition in Defense Technology.* Washington: U. S. Govt. Print. Off., 1990.

______. *Defending Secrets, Sharing Data.* Washington: U. S. Govt. Print. Off., 1987.

______. *Holding the Edge: Maintaining the Defense Technology Base.* Washington: U. S. Govt. Print. Off., 1989.

______. *Science, Technology, and the First Amendment.* Washington: U. S. Govt. Print. Off., 1988.

______. *Scientific Validity of Polygraph Testing: A Research Review and Evaluation.* Washington: U. S. Govt. Print. Off., 1983.

______. *Technology and East-West Trade.* Washington: U. S. Govt. Print. Off., 1979.

______. *Technology and East-West Trade: An Update.* Washington: U. S. Govt. Print. Off., 1983.

______. *Technology Transfer to the Middle East.* Washington: U. S. Govt. Print. Off., 1984.

______. *The Regulatory Environment for Science.* Washington: U. S. Govt. Print. Off., 1986.

Wallich, Paul, ed. *Managing the Flow of Technical Information: An Industry/Government Dialogue.* Sponsored by *IEEE Spectrum* Magazine. July 2, 1982. Washington, D. C.

AUTHOR INDEX

SUBJECT INDEX

C

D